new *french restaurant design*

新法国餐厅设计

ICI CONSULTANTS
法国亦西文化 编著

Direction 总企划
Chia-Ling CHIEN 简嘉玲

Documentation 资料收集与整理
Nicolas BRIZAULT

Chinese Translation 中文翻译
Shu-Chen LEE 李淑真

English Translation 英文翻译
Alison CULLIFORD

Graphic Design & Layout 美术设计与排版
Wijane NOREE

辽宁科学技术出版社

CONTENTS
目录

CAFE RESTAURANTS
酒吧–简餐–下午茶馆

new FRENCH RESTAURANT DESIGN
新法国餐厅设计

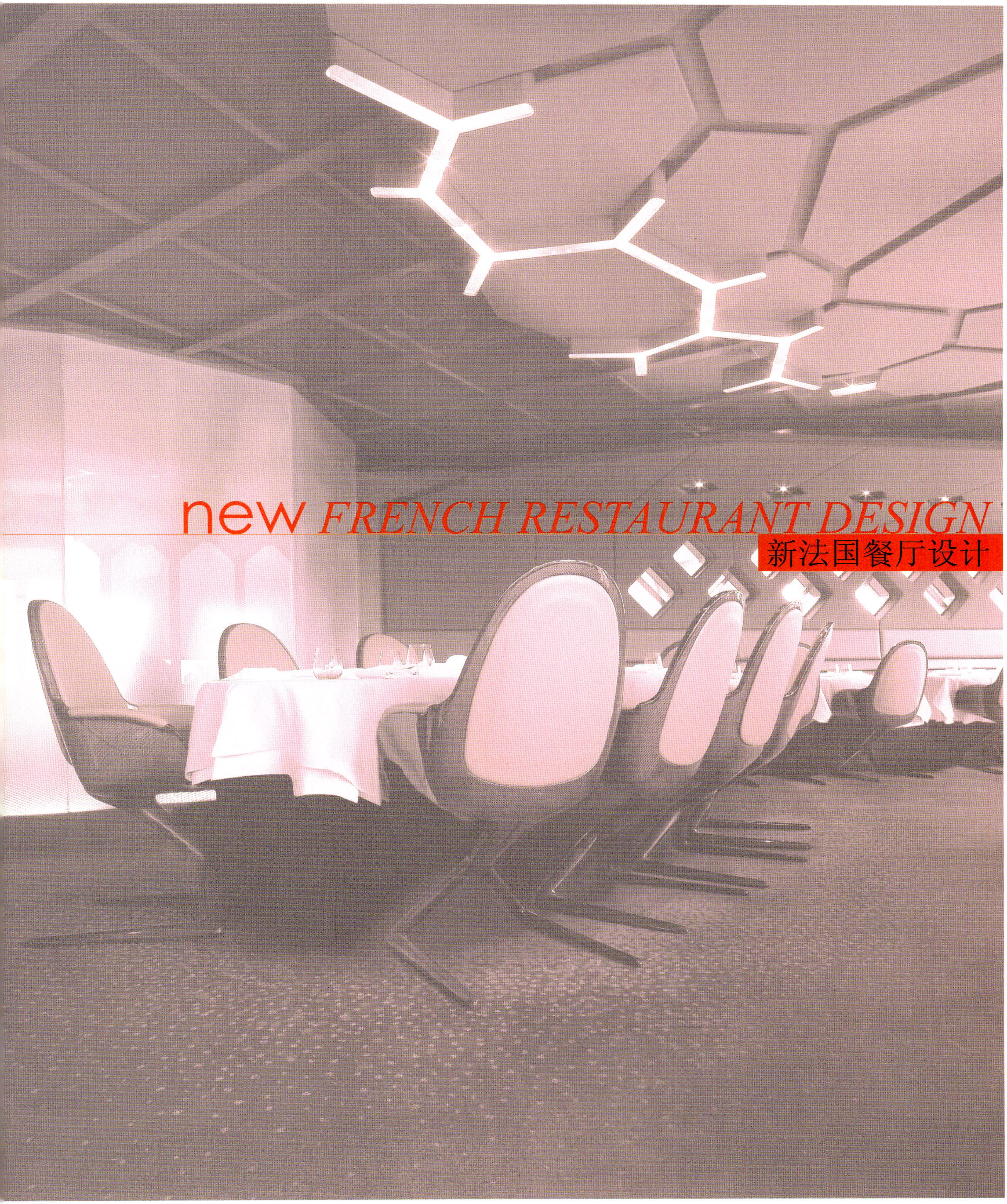

LOUNGE RESTAURANTS

休闲餐厅

设计师 Designer : Alexandre de Betak
地点 Location : Paris, France
完工日期 Completion Date : 2006
摄影师 Photographer : Bureau Betak

BLACK CALAVADOS

黑色卡尔瓦多斯酒吧餐厅

"We'll go to the Calavados tonight!" sang Serge Gainsbourg, celebrating the splendour of one of the most famous Parisian nightclubs from the 1950s up to the 1970s; the flower of show business at that time made this its hideout. The legendary bar-restaurant closed its doors at the end of the 90s. Today, transformed, rewritten for its own epoch, the Calavados has returned as an acronym: BC for Black Calavados.

This little theatre has been ideally conceived and put in the spotlight (or rather the shadows) by Alexandre de Betak, the artistic director who is the darling of the couture shows (Dior, Viktor & Rolf, Victoria's Secret, etc.). He was charged with designing the interior fit-out down to the smallest details. He has employed a completely modern radicalism here, breathing into the place the resolutely chic, intimate and rock spirit that was its own in the beginning.

The simple volumes of the restaurant are on a human scale. Betak here uses all the nuances of black, also working with the light in a very individual fashion. The furniture, also designed for the place, offers smooth, brilliant, matt or satin surfaces. Resin, black stainless steel, smoked mirrors answer each other in a space where everything has been done to deflect stares, or to see and be seen. Alexandre de Betak has revived the Black Calavados of 70s showbiz with the gymnastics of "BC".

"我们今晚去卡尔瓦多斯！"这是法国名歌手赛吉-甘斯布所唱的一句歌词，颂扬20世纪50年代至70年代间，巴黎最热门的夜总会卡尔瓦多斯有多么光彩夺目。此地成为当时演艺娱乐圈名人最爱出入的巢穴，然而这间成为传奇故事的酒吧餐厅于90年代末期结束营业。今日的卡尔瓦多斯以焕然一新的面貌和缩写名号BC（黑色卡尔瓦多斯）复出江湖，再度进入当代潮流。

这舞台般的餐饮空间是由亚力山大-德-贝塔克精心设计，他是高级服装走秀界（迪奥、维果罗夫、维多利亚的秘密等）最抢手的艺术总监，此次他负责整个内部的规划布置，直至最细微部分的设计。他采用十分现代的激进改变装修风格，给于餐厅绝然时尚摩登又亲密的氛围，且保有卡尔瓦多斯早年那种摇滚气息。

餐厅简洁的空间场域设计符合客人的舒适需求。贝塔克巧妙运用黑色各种色调，也以十分独特的方式设计灯光。为餐厅量身定做的家具呈现光滑、亮丽、雾面消光或上光的表面质感。树脂、黑色不锈钢和烟镜在空间中互相呼应反射，故意制造出令人眼花缭乱的效果，客人也可借机看看名人或秀秀自己。亚力山大-德-贝塔克再次将黑色卡尔瓦多斯推上犹如70年代娱乐圈那热门的舞台。

Plan
Restaurant

◇ BA13 DOUBLE PLAQUE SUR RAIL DOUBLE FACE
  + ISOLANT PHONIQUE
◇ REPRISE ENDUIT PLATRE

◇ STAFF
◇ BA13 DOUBLE PLAQUE MIXTE DOUBLE FACE
  STANDARD / HYDRO SUR RAIL + ISOLANT PHONIQUE

设计师 Designer : Olivier Gagnère
地点 Location : Paris, France
完工日期 Completion Date : 2007
摄影师 Photographer : Véronique Mati,
Philippe Schaff

15CENT15

15颂15 餐厅

In the heart of the Golden Triangle, the hotel Marignan houses a very contemporary and comfortable bar: the 15Cent15. Nathalie Richard, who wields the conductor's baton here, called on the designer Olivier Gagnère to give this place a warm and glamorous atmosphere, infused with modernism and a very Parisian elegance.

A play of materials, bringing together luxury and sensuality – velvets (Edmond Petit), leather, silky taffetas (Rubelli), a Taï Ping rug taking up the idea of a point de Hongrie parquet in a cocooning version – reinforce the "couture" spirit that is so beloved of this quarter.

This bijou bar, a real ice cream palace, places its decor, invites one to delight and incites one to sink in to the deep sofas of the neighbouring salons. Time stands still... The superb chandeliers of blown Murano glass (by the house of Veronese) emerge from a calisson adorned with gold leaf and magnify the curious objects sourced and assembled by the antiquarian and decorator Stéphane Olivier.

The chosen few will lose themselves in the astonishing, very British boudoir, on the borders of dandyism, which makes this place a confidential and unmissable address in the capital.

位在巴黎最热闹的金三角地带，马里涅索菲特酒店里有一间非常时尚舒适的酒吧：15颂15。酒店的总指挥娜塔丽-理查将名设计师奥利维耶-加涅尔请来负责此方案，希望他为此空间营造出温馨且魅力十足的气氛，也赋予它时尚感及花都巴黎的独特优雅气质。

设计师结合奢华与享受，善用各种高级质料的效果，如：爱德蒙-佩迪品牌的绒布、皮料、露倍利品牌丝般柔滑的塔夫绸，还有太平地毯集团所出产的地毯，其花样模仿匈牙利式斜条木地板的风格，但增添了毛绒绒的柔软质感。这些材质效果强调了某种"高级时装界"的气质，十分符合这个时尚区的精神。

吧台犹如一座名副其实的镜厅，闪耀亮丽的装饰，令人极欲靠过来畅然享受，也吸引客人走进位于一旁的厅房，舒适地躺在柔软宽大的沙发上。在此，时间似乎停止，维何内兹出品的穆哈诺水晶吊灯如此富丽堂皇，从内面饰了金箔的圆型结构中悬挂下来，柔和的灯光更美化了厅内的摆饰，这些奇特艺术品是由古董商兼装潢师史蒂凡-奥利维耶极其考究地搜集而来的。

首次前来的宾客很可能会迷失在那十分惊人、极其英式的小沙龙里，几乎接近纨绔主义的风格，而这也让此地成为首都巴黎一个只有内行人才知道、且绝不可错过的时髦酒吧。

设计师 Designer : Christophe Pillet
地点 Location : Paris, France
完工日期 Completion Date : 2004
摄影师 Photographer : Anoushka

BERTIE

贝尔帝餐厅

Situated in a pedestrian road just minutes from the Olympia concert hall, the Bertie takes its name from King Edward VII, whose real name was Albert, shortened to Bertie. Decorated by Christophe Pillet, the Bertie offers two different ambiances in a setting that is as contemporary as it is welcoming.

The restaurant, on the ground floor, has seating for 80 in a spirit where design comes to the fore, leaving its stamp in tones of grey, white and red. A large marble bar is perfect for enjoying a cocktail or a quick lunch. The simple and efficient character of this restaurant keeps the mood dynamic, while the large, white leather sofas invite guests to prolong the evening listening to the musical selection of the night: jazz, electro, trip hop or soul.

On first floor the space is more intimate while still being surprising and energetic, with the colour red lighting up the walls, and beige on the sofas, low tables and simple and welcoming armchairs that invite one to relax in a atmosphere that is 70s and thoroughly modern at the same time. The space can be "macro" with huge tables inviting lively get-togethers, or "micro", for tête-à-tête in a bright red semi-circle, under Bertie's suspended lights. The alcove lends itself equally to secrets and to showing-off, to being the centre of attention and to hiding away.

The green salon is invested with the lively and generous ambiance of a fresh and joyful green, contrasting with red reflected in one of the mirrors, which are so much a part of Bertie's. And as soon as the first rays of sun appear, the ground-floor terrace, protected from the bustle of the city, can seat around 80 guests.

贝尔帝餐厅位于奥林匹亚音乐厅附近的一条步行街上。贝尔帝是爱德华七世国王的别号，其真正的名字是爱伯特。冠以高尚皇家之名，贝尔帝餐厅由克里斯多夫-皮耶负责室内装修。在设计得既具当代感又十分温暖舒适的空间里营造出两种不同气氛。

位在一楼的餐厅空间可容纳80位客人，整体空间以出色的设计引人注目，且以灰、白或红色为色彩主调，彼此相得益彰。厅内设置了一座宽敞的大理石吧台，可以让人在此品尝鸡尾酒或使用一道午间简餐。此餐厅简洁有力的设计特色到了晚餐时刻仍保有迷人的活力；舒适的白色皮质大沙发令客人流连忘返，用餐后想继续留下来喝一杯，欣赏最新的精选音乐，陶醉在爵士乐、电子乐、神游迷幻乐或灵魂乐的乐声中。

相较之下，位在二楼的空间气氛显得更亲密雅致，但同样充满活力、令人惊奇。厅中的墙面饰以艳丽的红色，配上米色的沙发，摆设一些矮桌和设计简约舒适的座椅，让客人在这个带点70年代复古风，却又极其现代的气氛中轻松享受。这个空间可以很"宽阔大气"，摆置几张大桌，办场闹哄哄的聚会；也可以"小而亲密"，面对面坐在艳红的半圆形空间内。在贝尔帝的吊灯下。凹室空间的塑造，在此既成为展露台面、自我表现的空间，也可作为亲密对话、退避隐秘的角落。

绿色厅房为整个空间增添了愉悦、大方且诱人的酸甜味道，红色也透过墙上的一个镜面反射进来（贝尔帝餐厅内巧饰了众多镜子）。等春阳一到，远离着大街尘嚣的室外露天座，也可接待80位客人在此享用餐饮。

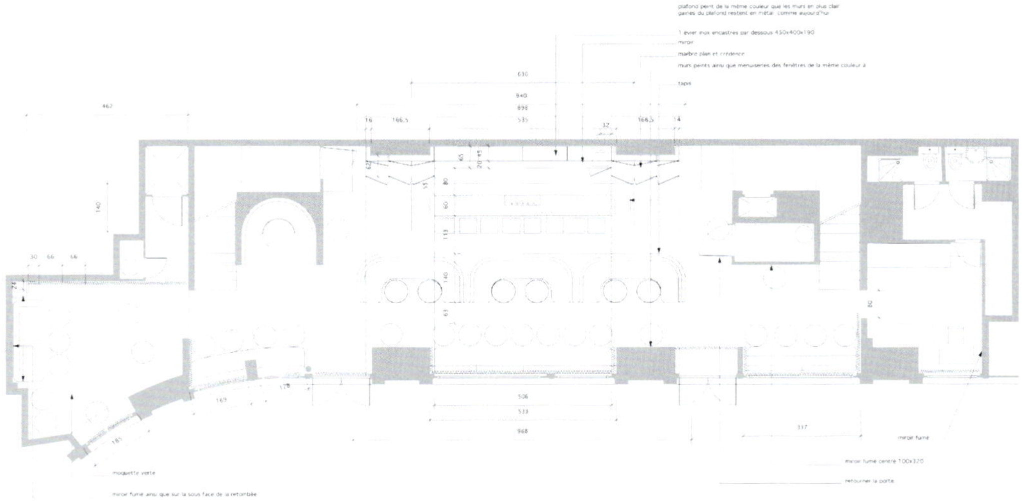

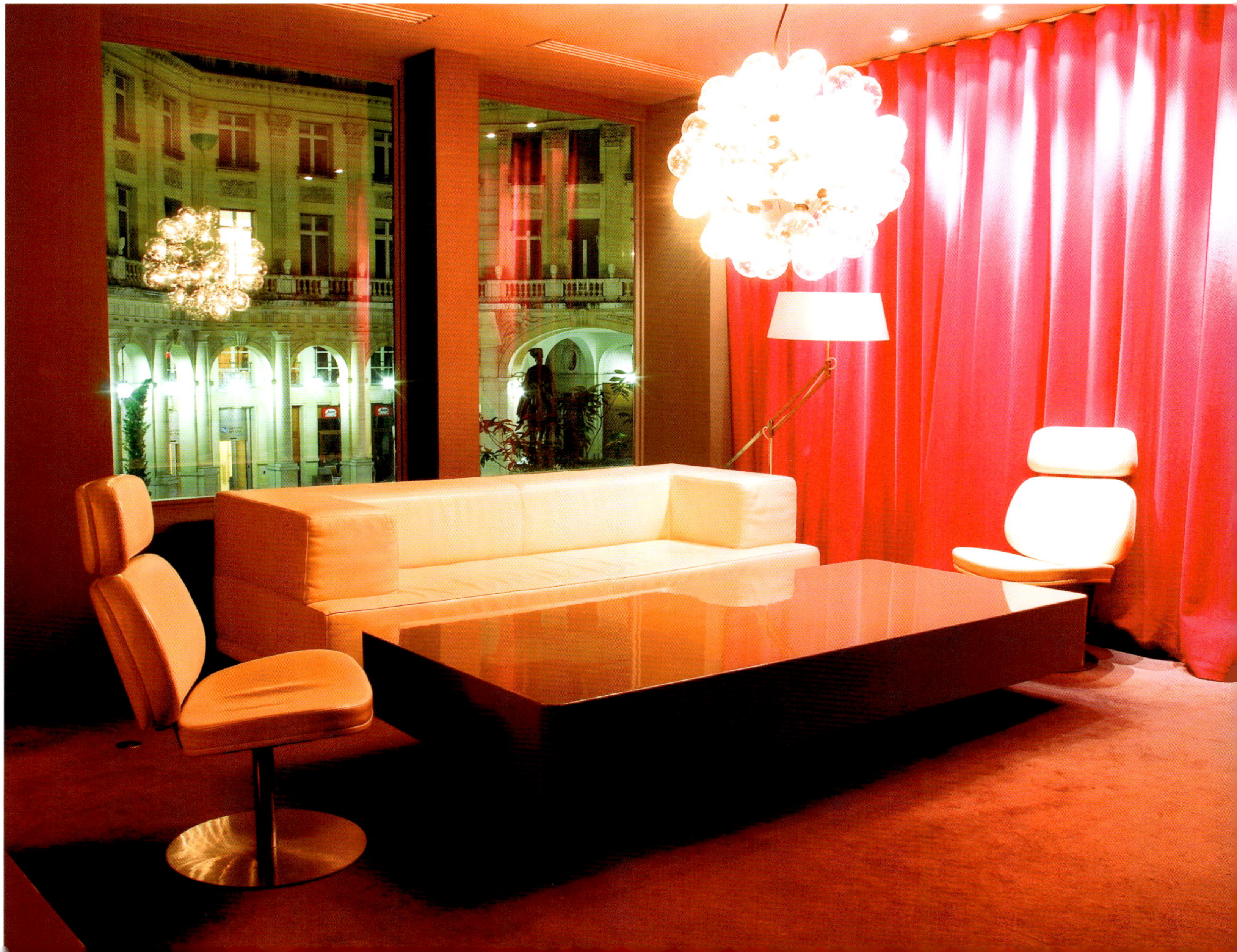

设计师 Designer : Gilles Le Gall &
Patrick Polianoff
地点 Location : Paris, France
完工日期 Completion Date : 2007
摄影师 Photographer : Wijane Noree

MINI PALAIS

迷你宫餐厅

In the heart of Paris, between the Champs-Elysées and the Seine, the colossal architecture of the Grand Palais spreads out, the symbol of a conquering 20th century. Built for the Universal Exhibition of 1900, it suffered serious structural damage over the years and had to be closed to the public in 1993. After a total renovation it was finally able to reopen its doors in 2007, and the idea of adjoining a restaurant was accepted. It would be installed on the Seine side, in place of the teaching rooms of the Architecture School.

The first intuition of the architects was to rediscover the scope and excess of the volumes, which a series of partitions and lowered ceilings had completely annihilated. The Ministry of Culture, in charge of the restructuring of the Grand Palais, wanted to create an ephemeral restaurant in a space of just two months, and partner was immediately found: Mini. This brought a style and a name, the "Mini Palais". From that point on, the architects used the off-beat and modern codes of the brand by contrasting excess and miniature, black and colours.

The room has been treated as a single volume with its eight metres of height, decorated with dark velvets, pierced with elegant vertical openings allowing one to glimpse the exhibitions installed in the large nave. On the other side, the architects have uncovered the high windows giving onto the peristyle and the Deglane colonnade which offers an astonishing view over the Seine and the Petit Palais. Immense coloured lights fall from a black ceiling, thus increasing the impression of height, and a collection of Baccarat crystal lamps, a nod to the Belle Epoque, illustrate the quirky side that the architects wanted to convey. Excess, originality, colour are the three notes of a project that brings together the architectural classicism of the place and the "terribly British" humour of the Mini brand.

此餐厅所在的大皇宫位于巴黎市中心，并且临近香榭丽舍大道与塞纳河，其庞大雄伟的建筑，成为辉煌20世纪的象征。宏伟的大皇宫建筑是为了1900年的世界博览会而建造的，历经岁月考验，而逐渐出现一些结构上的严重损坏，因此不得不于1993年停止对外开放。大皇宫经过全面的整修，终于在2007年重新开启大门，而为大皇宫添加一个餐厅的构想也在此时通过决议。餐厅被设置在建筑物内靠塞纳河的一侧，取代原来高等建筑学院的教室空间。

先前的一系列隔间设施以及被降低的天花板，完全破坏了空间原本的宽阔特色。建筑师们面对此方案的第一个直觉就是为此地重新找回一种超尺度的广阔体量。负责改造大皇宫的法国文化部希望能快速地在两个月内开创一个灵活变动的餐饮空间；他们很快找到了一个合作伙伴："迷你国际车业"。此项合作立即为餐厅带来一种属于"迷你"的风格，并且带来一个名字："迷你宫"。依此为出发点，建筑师们采用迷你品牌独特的现代风格，巧妙运用超大与缩小、黑色与多彩之间的对比来进行设计。

建筑师将餐厅设计成天花板高达8米的单一宽阔空间，厅内铺饰深色绒布，临靠大殿这一侧的墙面开凿了几扇优雅的长方形玻璃窗口，使客人可以透过这些窗口欣赏在大殿中的展览。建筑师在餐厅另一侧则重新设置一扇扇原有的高大窗户，让人得以望向窗外雄伟的列柱廊，再望出去，便是小皇宫和塞纳河那一片扣人心弦的美景。一盏盏彩色的圆形巨大灯饰从黑色天花板悬吊下来，更增强了天花板高阔的感觉；一系列高低错落的巴卡拉水晶吊灯，表现出欧洲美丽年代的风华，也呈现出建筑师们想诠释的那种将不同时代风格并置的美感。此设计方案正是运用"超尺度"、"时空落差"和"色彩"这三个特点来将一个古典的建筑场所与迷你品牌那种极其英式的幽默结合在一起。

设计师 Designer : Christian Ghion
地点 Location : Tokyo, Japan
完工日期 Completion Date : 2007
摄影师 Photographer : Christian Ghion

LA PATISSERIE PIERRE GAGNAIRE

皮埃尔-贾内甜点店

"La Pâtisserie Pierre Gagnaire", located on the 4th floor of Japanese department store Takashimaya in the lively district of Shibuya, Tokyo, opened its doors in December 2007. People can taste a selection of cakes, biscuits and fruit salads, displayed like gems in the windows of a jeweller's.

Christian Ghion has created a very discreet and charming world for the shop: the wooden panelling and cream leather seating standing out against the autumn colours of a "Bayadère" carpet in plum, chocolate and khaki, brightened by thin lines of light blue and orange. Elsewhere, the enveloping, soft lighting remains dynamic, and is given a cadence by small rectangles of white fabric hung on straight and geometric lines around the lamps.

Finally, a large expanse of pistachio- and coffee-coloured wallpaper imitating damask fabric conceals in its motif an amusing nod to that famous French cake, the "religieuse". A French spirit and Japanese culture combine and meld here, providing an extra pleasure for visitors to this pâtisserie celebrating the talent of Pierre Gagnaire.

皮埃尔-贾内甜点店于2007年12月开张，该店位于日本新宿高岛屋百货公司的四楼，地处东京热闹非凡的涉谷区。店里可品尝到一系列精致的糕点、油酥饼和水果沙拉，一款款甜点摆饰得像是高级珠宝店橱窗里的闪亮珠宝。

设计师克里斯提安-吉翁在此构思出一个拥有含蓄魅力的世界：墙面覆盖了色调深沉却温馨的木材，米白色的皮质软垫沙发与亮丽的地毯形成强烈对比。这个巴亚德直条纹地毯采用了秋天的色调，深紫、巧克力色或卡其色条纹，夹上几条细绵棉的天蓝色和橙色，使整片地毯的色调更亮眼。此外，店内灯光几乎像绒毛般柔和，却又显得活力十足，加上从天花板垂吊下来的白色长方形小布块，在灯光的四周以直线呈几何形方式排列开来，为整体灯光氛围增加了韵律感。

一道装贴了壁纸的主要墙面采用了淡黄绿色配咖啡色，并模仿花级的感觉，其花纹图案让人看了不禁联想到一种非常道地的法式甜点：双球包奶油蛋糕。此甜点店不仅让人见识到皮埃尔-贾内这位名厨的才艺，又能使人领略到法国时尚风格与日本文化相遇相融合的成果。

La Patisserie
PIERRE GAGNAIRE

设计师 Designer : Anégil
地点 Location : Paris, France
完工日期 Completion Date : 2007
摄影师 Photographer : Xavier Béjot

LES GRANDES MARCHES

大阶梯餐厅

Les Grandes Marches ("The Grand Staircase"), whose history, surroundings and location have made it a Parisian legend, has taken on a new life as a multi-faceted bar-restaurant. As befits its age, the place has lived through several changes, and is proud of this heritage. The setting is thus a reflection of our age: a timeless patchwork where year, fashions, materials and moods are juxtaposed.

In the purest bistro tradition, a large, pewter-topped bar contrasts with a forest of lampshades in coppery colours. The mixture of furniture is reflected in enormous, mirror-polished copper spheres. A series of alcoves upholstered in shimmering, wheat-coloured fabrics dominates the space, while offering a "view" that takes us back to the Bastille quarter at the dawn of the 20th century.

The sweeping staircase has steps clad in a soft orange carpet, while two chandeliers made up of clusters of tubular glass illuminate the centre of the room. Lamps casting hypnotic shadows reinforce the throwback to an earlier era. The walls have been decorated with a lightly combed, coppery paint effect, and coloured velvets.

Hidden away from curious stares is an intimate bar with a distinctly "hairy" look: the moumoute bar. The bar, the lampshades, the stools and the armchairs are all furry. In this silvery setting, five lounging alcoves allow one to enjoy a drink stretched out as if on a bed, while admiring the panoramic view of the place de la Bastille.

大阶梯餐厅是个传奇场所，然而这个特色不仅缘于它的历史，也来自于它的环境及其地理位置。今日的大阶梯餐厅则是在这样一个多样面貌的空间中蜕变重生。此餐饮空间一向跟随着时代潮流演变，因此经历过多次装修改变，此丰富的时间遗产成为餐厅的一种骄傲，它的处境正反映出我们这个时代的样貌：不同年代、各种潮流、多样材质以及多元的氛围在此汇集，最终呈现出一种超越时空的复合风尚。

跟昔日最传统的酒馆一样，厅内有一座锡质边饰的大吧台，与吧台上方那一片金铜色的灯罩产生对比。风格多样的家具映照在一盏盏尺寸惊人、具有圆滑镜面的铜质球形灯罩上。呈麦穗状排列的一系列箱式隔间座位运用了绚丽多彩的布料围塑出亲密的小空间，同时借助墙上的影像作品来让人欣赏巴黎巴士底狱街区上个世纪初的景象。

餐厅中央陈设着一座大楼梯以连接上下楼层，阶梯上铺着柔软的橙色地毯，两盏管形玻璃聚制而成的流苏状吊灯照亮了整个餐厅中央。楼上带着催眠般光影的灯光，更加强了时代错置的感觉。墙上则采用了具有铜褐色泽与细微纹路的材质，也配合色彩迷人的绒布。

餐厅里还有一个较隐密的空间，一个"毛茸茸"的亲密酒吧区：毛毛酒吧。吧台、灯罩、高凳和扶手椅都是毛茸茸的。在这带点银色调、珠宝盒般的空间中有五个通铺式厢房，客人可躺下来喝一杯，同时欣赏窗外的今日巴士底狱广场全景。

设计师 Designer : Gérard Olouman
地点 Location : Paris, France
完工日期 Completion Date : 2005
摄影师 Photographer : Wijane Noree, Royce

ROYCE

莱斯餐厅

First of all you encounter the unique boudoir of the Royce, then descend a few steps and the smooth and festive world of the lounge begins. Organized around a glass bar resembling the prow of a ship, the Royce plays with moving light that ripples between turquoise, yellow, pink, orange... In the Fragonard salon visitors can also sink into large leather sofas. The light is filtered by a curtain of steel beads, preserving an atmosphere of intimacy.

With its terrace opening onto a Japanese garden, the restaurant of the Royce immediately projects a cosy and discreet ambiance. A glass wall hiding a giant TV screen, linen curtains and soft seats in Alcantara are softened by the natural light from the garden, which highlights the wenge wood furniture.

The appetite is stimulated by cream and chocolate tones, discreetly enhanced by the globe lights here and there and the glimmers from the fireplace. Then, on a starry night, the terrace opens up on a Japanese garden in the heart of Paris! An oasis of fresh air, it is full of magic with its fountains evoking the Far East, and this ambiance is also found in the interior with its multiple representations of Buddha in the form of statues and surprising photographs. Relaxation and escape, soothed by the gentle sound of water!

一进到莱斯餐厅，首先看到的是风格独一无二的沙发座式精品店，然后走几个阶梯下去，便踏进了一个温暖舒适、充满热闹气氛的沙发酒吧世界！整体空间以形如船首般的玻璃吧台为中心来设计，并且散发着充满动感的灯光，像海浪一般，在青绿、黄色、粉红、橙橘色之间翻转。在此，访客可享受名牌香水法拉贡那尔专属沙龙和真皮大沙发。柔和的灯光轻轻透过钢珠垂帘，营造出一个安适的亲密空间。

莱斯餐厅的露天座位区面对着一座日式庭院，更强化了餐厅温馨私密的特殊氛围。花园里的阳光总是如此自然地进入室内，照映在遮着大型荧幕的玻璃墙上，照暖了亚麻布帘和豪华复合面料制的柔软座椅，也照亮了非洲刚果红木制的家具。

餐厅内以奶油色和巧克力色搭配的主色调引人胃口大开，而室内四处回光点点和壁炉火光的光影交错，更衬托出这个色调的优美。在满天星空的夜晚，露天餐饮区对外开放，使人们深处花都巴黎，却又能观赏日式庭院！人们在此呼吸新鲜空气，在一座座令人联想到东方氛围的喷泉旁，享受露天座所散发出的特殊魔力。餐厅室内也摆饰了诸多佛像雕塑以及令人惊叹的摄影作品，让人们在这个充满异国情调的场所可以放松享受，并在细水低回之中，任时光摇曳。

设计师 Designer : Jacques Garcia
地点 Location : Paris, France
完工日期 Completion Date : 2006
摄影师 Photographer : Eric Cuvellier,
Patrick Aufauvre, Wijane Noree

LE FIRST

第一餐厅

"le First, restaurant boudoir paris" is located at number 234 on the famous rue de Rivoli. Designed by Jacques Garcia, one of the most renowned interior designers worldwide, and orchestrated by chef Gilles Grasteau, le First invites its guests to taste authentic and inventive cuisine in an elegant setting. Under the arcades of the hotel The Westin Paris in the 1st arrondissement, le First offers a contemporary decor with a baroque touch. It is truly a place where design and gastronomy are bound together.

Jacques Garcia was given the mission of designing the restaurant in a unique style that creates a timeless bridge with history. Foliage and golden lances are printed on the fitted carpet as an echo of the Tuileries Gardens, whose branches can be seen swaying through the windows. A skilful set of mirrors and curtains play with the light, giving the setting the unique atmosphere of a contemporary boudoir.

This new "place to be" in central Paris reflects the passion for art shared by Jacques Garcia and Gilles Grasteau. It praises design and gastronomy in a subtle fusion. The cuisine remains simple but inventive, and the design is sober yet exceptional, bringing together purple, velvet and silks. The space, divided by a row of pilasters that structures the place on two levels, is elegant but relaxed.

"le First, restaurant boudoir paris" also welcomes l'Instant Boudoir, a tea-room, and, when the weather permits, le First extends outdoors to become a haven of dining tranquillity – La Terrasse.

"第一餐厅"位于巴黎第一区，在极为著名的里沃利路234号，由国际知名的室内设计师贾克-加尔夏设计，并由吉勒-格拉斯多大厨主掌烹调大任，盛邀宾客在一个高雅的环境空间中品尝既地道又充满创意的美食。这个处于威斯汀酒店的拱廊建筑下的"第一餐厅"，是一个微带有巴洛克风格却极具现代感的室内空间，并将设计与美食紧密结合在一起。

贾克-加尔夏的设计任务是要为餐厅创造一个独特的风格，成为一座超越时空的历史桥梁。厅内地毯上印饰叶子和金色长矛的图案，仿佛是窗外杜勒丽花园一幕幕美景的投映。室内的镜饰与窗帘彼此巧妙搭配，并与灯光效果产生极佳呼应，赋予室内一种时尚沙龙的独特氛围。

这个绝不能错过的巴黎新餐厅反映出贾克-加尔夏和吉勒-格拉斯多两位大师对艺术的共同热情，也是室内设计与美食融为一体的巧妙结晶：餐厅所提供的餐点虽然样式简单却极具创意；而室内设计则是在含蓄之中带有独特性格。这个结合紫色、天鹅绒和丝绸的空间，以一列长方柱将餐厅分构成两种层次，整体显得十分高雅却也轻松舒适。

此餐厅还设有一个名为"沙龙时刻"的饮茶沙龙，而且当天候允许，餐厅也开放室外的"露天座"，成为一个享受宁静晚餐的天堂。

设计师 Designer : Jacques Garcia
地点 Location : Paris, France
完工日期 Completion Date : 2007
摄影师 Photographer : Anne-Laure Jacquart

LE RESTAURANT

"餐厅"

Le Restaurant. This simple name is enough. Once past the doors of L'Hôtel, this old Parisian pavilion raised by six stories during the Directoire to become a precursor of the hotel, the visitor enters into a tasteful world. The deep chairs upholstered in velvet invite him to temptation and tasting, in this magical place only a few steps from the Ecole des Beaux-arts.

The atmosphere of Le Restaurant is, however, without ostentation or excess, as perfectionism demands. This creation is naturally the work of one of the world's greatest ambassadors of luxury, Jacques Garcia. His design plunges Le Restaurant into the spirit of a private salon, under a huge glass roof in the Eiffel style. The space opens on to a patio, particularly pleasant in fine weather, with its fountain, a masterpiece by the great 18th-century architect Claude-Nicolas Ledoux.

Epochs mix together here, as well as the lines – straight and serious, or airy with perfectly drawn curves – and warm and cold colours, everything coming together to create a perfect scene, in the spirit of L'Hôtel itself.

"餐厅"，这简单明了的店名已足够。如同它所在的酒店，名称就叫"酒店"，这栋古老的巴黎楼馆于法国督政府时代加盖了六层楼，并成为接待住宿的地方。一跨进酒店大门，宾客即进入一个崇尚品位的世界。覆有绒布的深椅座盛邀着客人，在这个临近高等艺术学院的神奇之地享受奢华诱惑与品尝美馔。

餐厅装修虽然华丽，其所呈现的氛围却不显得浮夸与过度，整体设计是严谨与完美的典范。这精典设计自然要归功于贾克-加尔夏，这位透过空间设计在世界各地传播华贵之美的使者。他的设计使"餐厅"在具有埃菲尔建筑风格的巨大玻璃天棚下，沉浸在一种贵宾沙龙的气氛中。餐厅空间朝一个内庭开展，在阳光时节里更显得舒适怡人：内庭里饰有一座喷泉，是18世纪伟大建筑师勒都的杰作。

在此餐厅，各时代的装饰风格融汇一堂，也结合了各种形体线条：有严肃笔直的线条，也有较为轻盈飘动或完美勾画的圆弧线条；此外，冷暖色调也相间融合。这一切设计构思皆为了塑造出精心而完善安排的室内场景，以与"酒店"的整体精神相符。

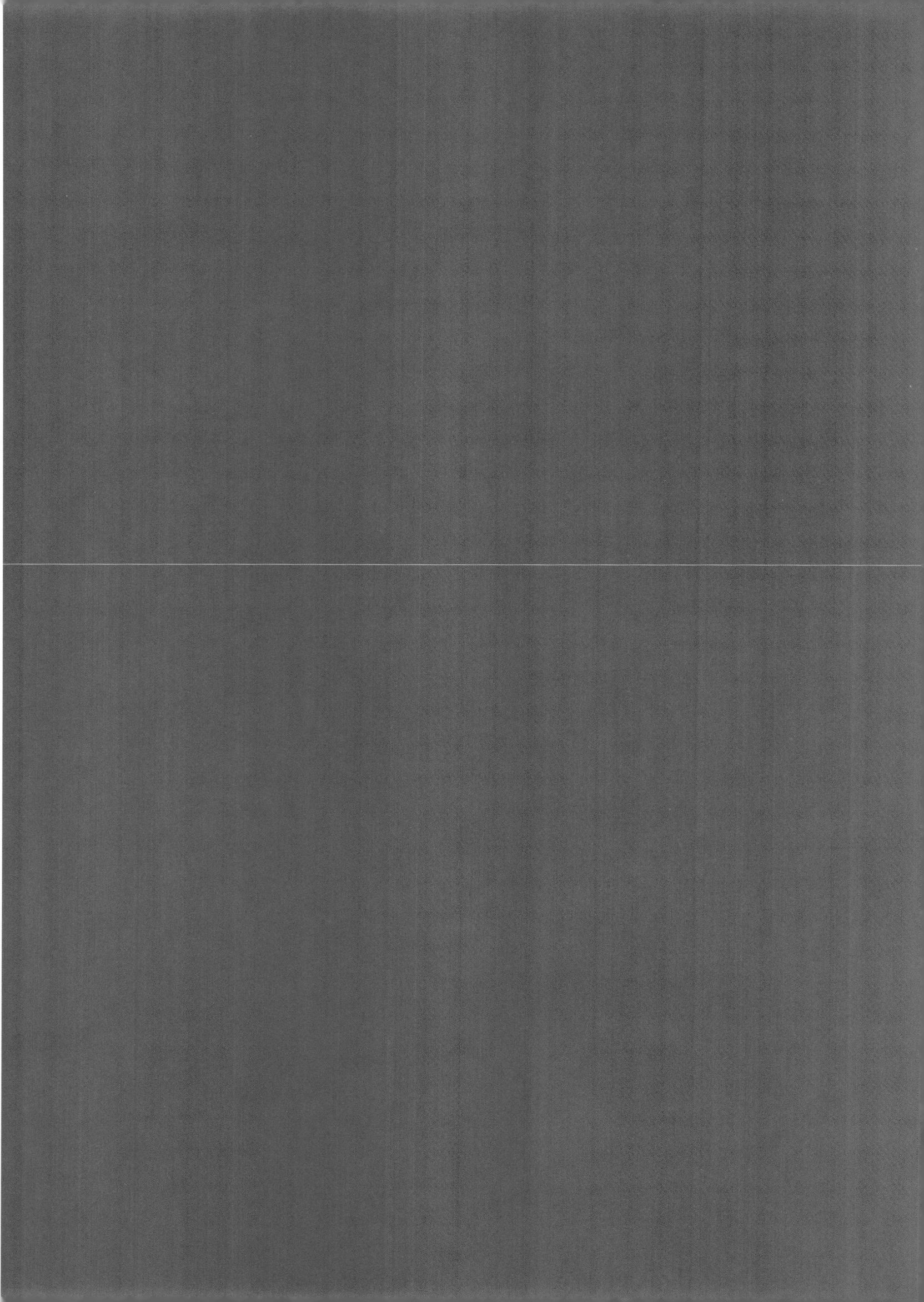

FULL SIT-DOWN RESTAURANTS
风尚餐厅

设计师 Designer : Christian Biecher
地点 Location : Paris, France
完工日期 Completion Date : 2008
摄影师 Photographer : Luc Boegly

FAUCHON

馥颂餐厅

For this place which evokes Paris and its multiple facets, the architect Christian Biecher was inspired by the codes of the Fauchon brand: the historic pale pink, which is the most important colour to Fauchon, with its sweet and appetizing tone; a black and white graphic, and also a grape motif. He decided to bring them together in an environment of very strong light that is nevertheless comfortable, perfectly adapted and conceived down to the last detail.

The interior fit-out of the 300 m² restaurant rests on the play of reflections, on the transformation of the atmosphere according to the time of day and the changes of light. For that, another brilliant colour, silver, has been added and envelops the space, redraws it and displays it through a wall of mirrors, curtains of glass beads and leather sofas, which are also silver. The omnipresence of this very metallic spirit is softened by a grey carpet whose pattern depicts generous and peaceful bunches of grapes.

The space therefore revolves around a discreet softness through which runs a revitalizing seam, adapting perfectly to the identity of Fauchon itself.

为了设计这代表巴黎精品时尚及巴黎多样化面貌的场所，建筑师克里斯提安-毕谢的灵感取自馥颂品牌最具象征性的符码：粉雾桃红色（色彩在馥颂品牌中扮演着极其重要的角色，尤其是这个历史性的粉雾桃红色，为产品带入温柔及高贵美味的形象）、黑白文字设计、甚或还有葡萄图案。建筑师将这些元素融入一个灯光极为强烈却拥有舒适氛围的空间中，连最微小的细节都巧妙融入了这些设计符码。

咖啡坊300平方米的室内空间设计主要在营造镜面反射和材质反光效果，以及跟随时辰变化的气氛和多彩多姿的灯光照明。为此，另一个拥有灿烂光芒的色彩：银色，也加入设计行列，涵盖整个空间，勾画出另一种空间感。银色透过一面镜墙、玻璃帘幕和皮沙发而铺展开来。整个空间中处处可见这种极具金属质感的元素，但搭配着饰有大方恬适的串串葡萄图案的灰色地毯，整体又显得温柔多了。

餐厅空间就在这种既含蓄柔和又绝然充满活力的精神中开展铺陈，完全符合馥颂品牌的形象。

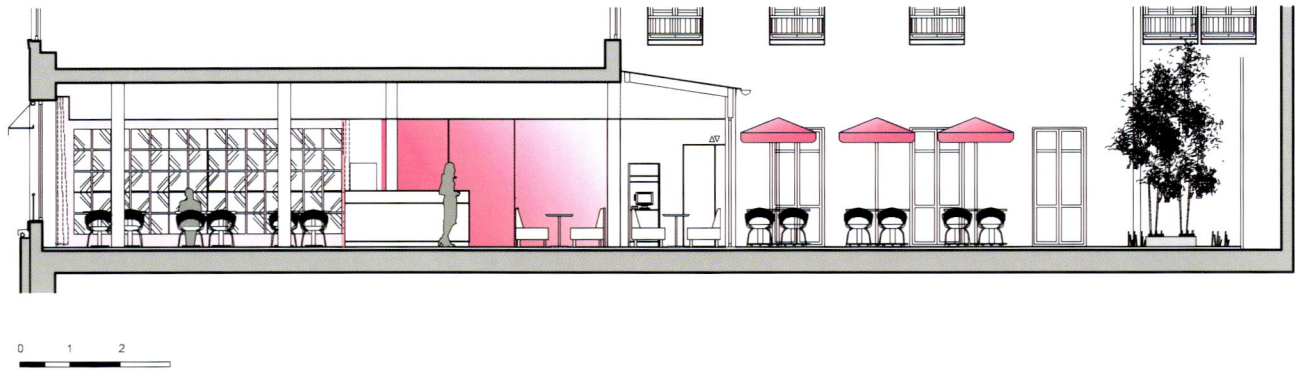

0 1 2

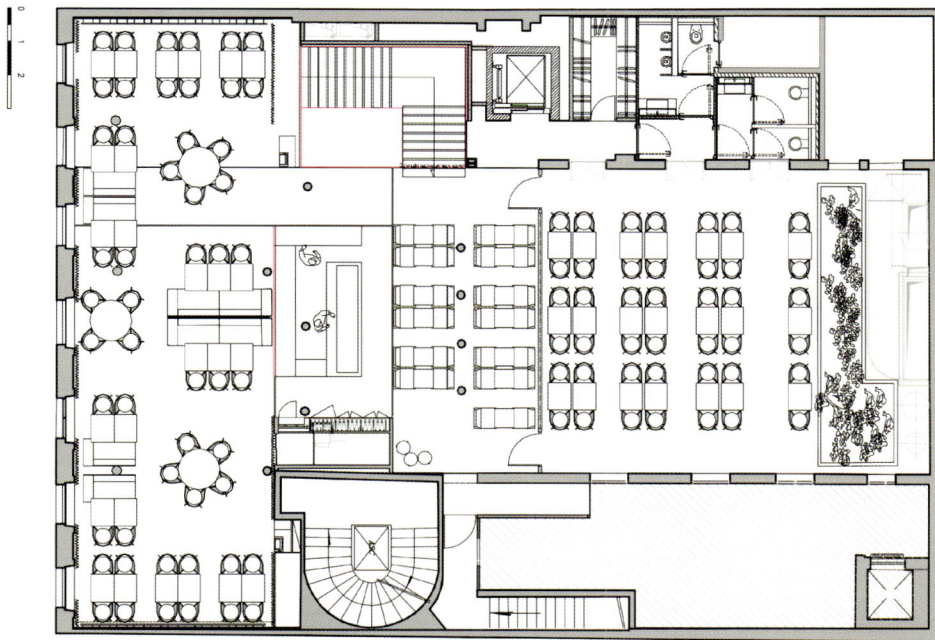

设计师 Designer : Patrick Jouin
地点 Location : Paris, France
完工日期 Completion Date : 2007
摄影师 Photographer : Eric Laignel

LE JULES VERNE

儒勒-凡尔纳餐厅

"In Gustave Eiffel's day, going up in a high building was far from being an everyday experience. It was one that provoked emotion. As for the construction of the Tower itself, it was a real technical feat, a truly audacious work of engineering. That is what inspired my work on the Jules Verne. I imagined a scene set for emotion, made up of several focal points and an imaginative decor. At the same time I wanted to integrate modern techniques that could also create beauty." At the Jules Verne, Patrick Jouin has created a contemporary decor where the space allows the eye to circulate. By day, natural light invades the restaurant and illuminates it. At night, the atmosphere is more intimate and sensual.

A luminous glass wall at the heart of the restaurant provides the strongest element of the visual plan of the "new Jules Verne". It draws a circle around the kitchen, thus creating a "moving pantry" behind which the waiters circulate. From the dining area, the aluminium honeycomb that the partition contains allows one to watch the changing shadows of the waiters and to see the energy that emanates from this mysterious bustle of activity. One can also see the oblique lines that perfectly reflect the architecture of the monument but which are none other than the service shelves of the pantry. This partition, which diffuses a changing light, offers a counterpoint to the panorama by extending the horizon line.

In the setting of the Jules Verne, technology and elegance mix harmoniously. A perfect illustration is the chair designed by Patrick Jouin. Bringing together leather and carbon fibre, it is a fluid and light object whose development took eight months of work. As comfortable as the chairs are the leather banquettes that face them. These have their own "sky", the whole construction forming an enveloping cocoon.

"在古斯塔夫-埃菲尔的时代，往高度发展并不是件稀松平常的事，那是会引发强烈情绪、令人感动的经验。至于铁塔本身的建造，那更是一件技术上极为了不起的壮举，一件极其大胆的工程杰作。这给了我设计儒勒-凡尔纳餐厅方案的灵感。我想象一出由许多高潮迭起的时刻组成的感人剧情和一个会引人织造绮丽梦想的布景：同时，融入今日的技术，这些高科技也可以创造出美感。"帕特里克-朱伊恩为此餐厅设计出十分当代的装修空间，让视线能够自由流转。白天，自然光倾泻进来，照亮整个餐厅；晚间，气氛变得较柔和较感性。

餐厅中央有座玻璃制的光亮环形墙，构成"新儒勒-凡尔纳"视觉装置上极为强势的元素。这面隔墙绕着厨房形成一个圆形环墙，也因此框造出一个"旋转后场"，在墙后面便是工作人员忙进忙出的地方。玻璃隔墙掩住的铝质蜂窝结构让坐在厅内的客人能够看见厨房里影影绰绰的动态，并感受到那神秘的忙碌工作所散发出来的活力。客人也会隐约看见一根根的斜柱，跟铁塔本身的建筑产生完美的对应，然而这些斜面结构只不过是后场配膳室的置物架。这座玻璃隔墙散发着变化的光线，与延伸地平线所望出去的全景形成巧妙对称。

在儒勒-凡尔纳这闪亮的餐饮空间里，高科技与高雅气氛和谐地交集共存，由帕特里克-朱伊恩设计的座椅便是一个完美的例子：它们结合了皮质与碳纤维，呈现出具流线感又轻盈的家具，这座椅的制造不易，光是研发工作就花了八个月的时间。有了舒适的扶手椅，摆置在对面的皮质长形软垫沙发也以一样的舒适相回应，延伸出去便是一片无垠的长空。整个餐厅的格局摆设令人感觉犹如被包裹在蚕茧中一样地温暖适意。

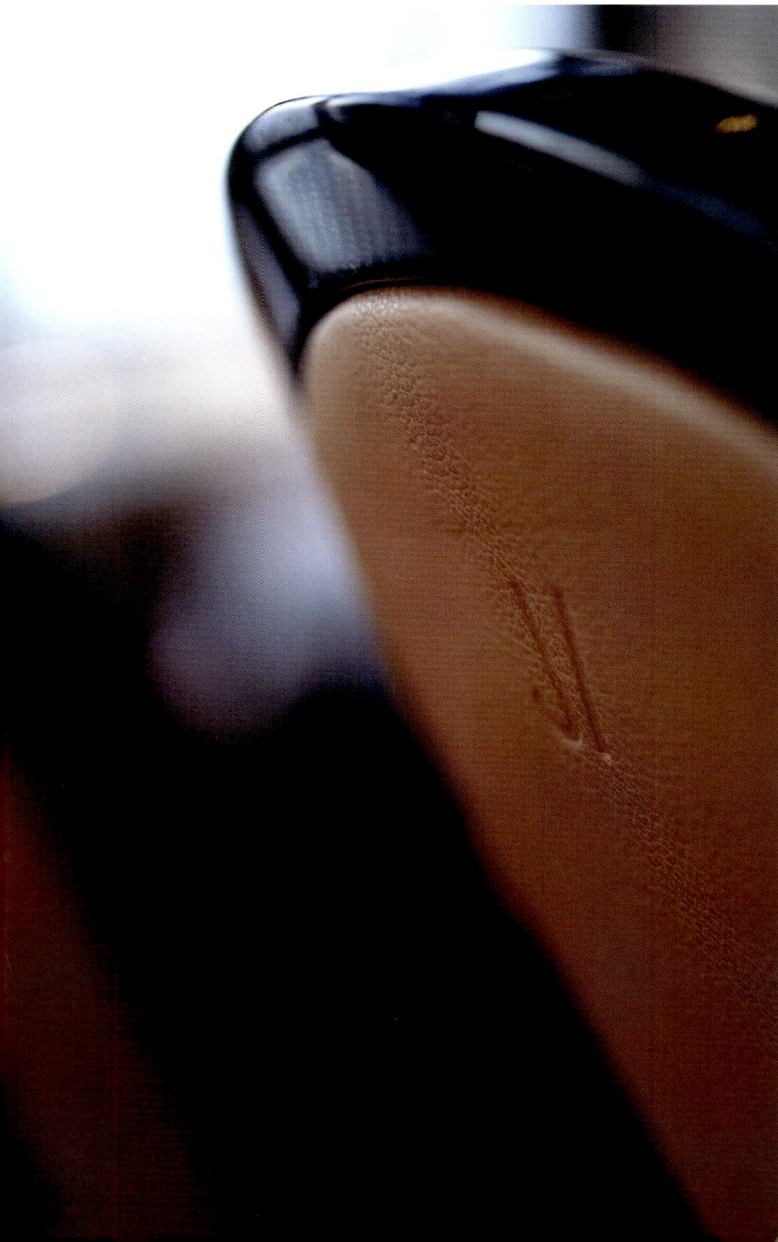

设计师 Designer : Donatelle Piana &
Philippe Batifoulier
地点 Location : Lyon, France
完工日期 Completion Date : 2008
摄影师 Photographer : Eric Saillet

LE BOUDOIR

不朵儿餐厅

The design of the Boudoir is based on two colours, red and black; two wall materials, MDF "curtains" and black Perspex; two types of lighting, candles and digitally controlled lights; two veiled references, the ceramic tiles with a monograph design on the floor and the beaten iron of the baobabs – all this to bring out two ambiances: day and night.

Serious attention has been paid to the lighting, allowing the Boudoir to metamorphose throughout the day. Through sophisticated programming, the lighting creates specific ambiances for the very chic lunchtime service, the aperitif and lounge restaurant lounge service in the evening, and finally the nightclub version.

By day, pink light gently plays with the draping of the walls, keeping it sensible. In the evening, this pink puts on the rouge to become one element in the setting of a spectacular scene. The hanging lights are turned out, the majestic ceiling lights are turned down and become the theatre for a festive play of lights.

The forms of the furniture and the bar are gentle and languid, like the armchairs whose curves are inspired by the back of a woman, and like the Corian bar top, which is rounded at each end. Finally, poetic areas have been placed under the baobabs created by Madagascan craftsmen, their fine boughs accentuated by faint LED lights.

不朵儿餐厅的设计构思透过两个色彩(红与黑)、两种墙面材质(有褶裥的布帘与黑色有机玻璃)、两款灯光(烛光与资讯化照明设备)以及两个巧思(戳印在地上的陶土与锻铁面包树)来表现,这一切都为了营造出两种时空气氛:日与夜。

精心设计的照明系统使餐厅依着时辰化身转变,经由一套十分先进的程式设定,让灯光的变换营造出各种特殊的气氛:午餐的商业时尚氛围、晚餐前的轻松光线、晚间沙发酒吧餐厅的柔和灯光,还有舞厅式疯狂的五光十色。

在日间时段乖巧地照在墙面布帘上的粉红灯光,到了夜晚却突然涨红,变成充满戏剧性的舞台灯光。而此时,吊灯熄灭,巨大的灯罩降下来,成为彩光舞影的屏幕。

桌椅和吧台的形体设计都采用温和柔滑的线条,例如扶手椅的设计,其曲线极像性感女郎的背脊,还有,吧台的可丽耐材质台面两端也呈圆弧形。此外,设计师在由马达加斯加岛的工匠制作成的锻铁面包树下,安置了几个诗情画意区,此处微弱的荧屏光线衬托出树上细长的枝叶。

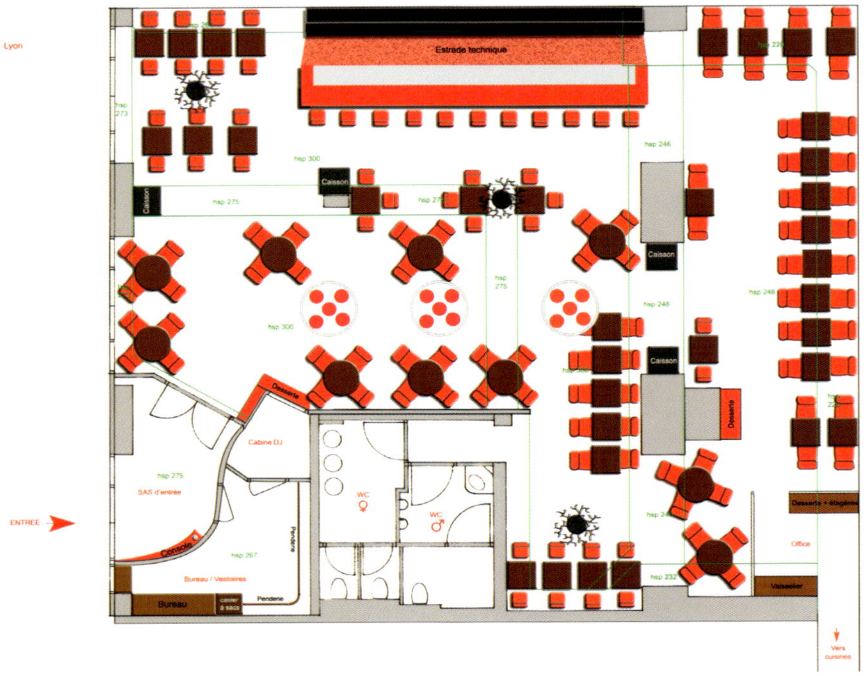
Lyon
Estrade technique
Caisson
Cabine DJ
SAS d'entrée
ENTREE
Bureau / Vestiaires
Penderie
Bureau
WC
Desserts
Desserts + étagères
Office
Vaisselier
Vers cuisines

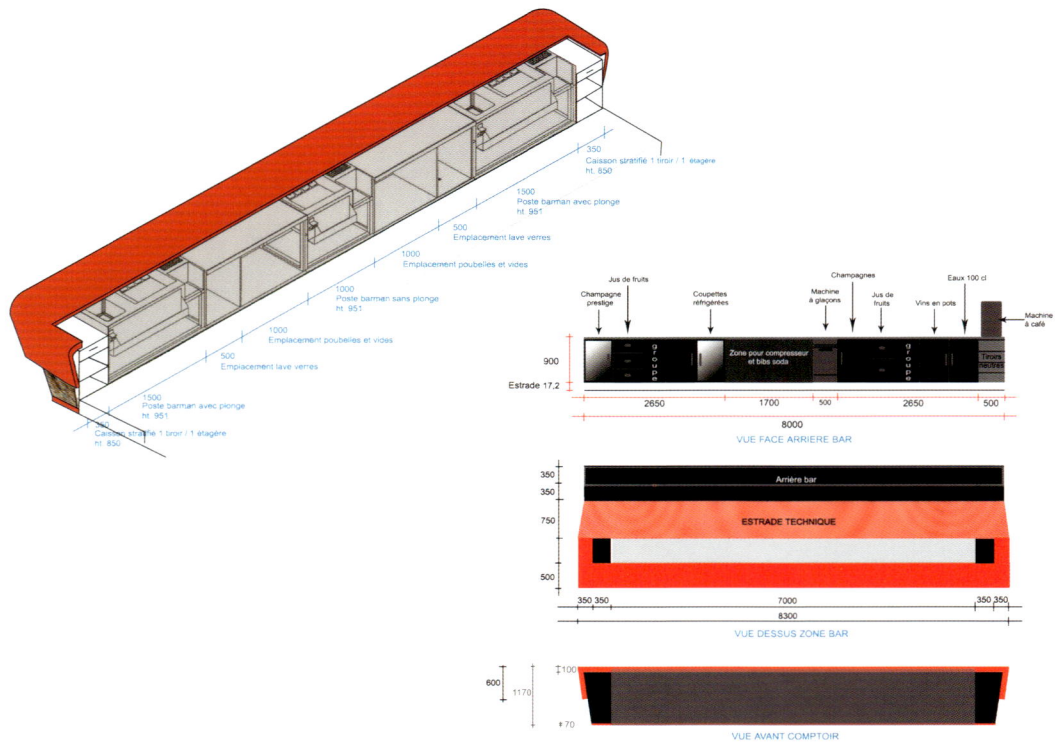

350
Caisson stratifié 1 tiroir / 1 étagère
ht. 850

1500
Poste barman avec plonge
ht. 991

500
Emplacement lave verres

1000
Emplacement poubelles et vides

1000
Poste barman sans plonge
ht. 951

1000
Emplacement poubelles et vides

500
Emplacement lave verres

1500
Poste barman avec plonge
ht. 991

Caisson stratifié 1 tiroir / 1 étagère
ht. 850

Jus de fruits

Champagnes

Eaux 100 cl

Champagne
prestige

Coupettes
réfrigérées

Machine
à glaçons

Jus de
fruits

Vins en pots

Machine
à café

900

groupe

Zone pour compresseur
et tibs soda

groupe

Tiroirs

Estrade 17,2

2650   1700   500   2650   500

8000

VUE FACE ARRIERE BAR

350

Arrière bar

350

750   ESTRADE TECHNIQUE

500

350 350   7000   350 350

8300

VUE DESSUS ZONE BAR

100

600   1170

+ 70

VUE AVANT COMPTOIR

设计师 Designer : Olivier Gossart
地点 Location : Paris, France
完工日期 Completion Date : 2006
摄影师 Photographer : Osmose

OSMOSE

欧思摩兹餐厅

Situated just a few steps from the "Maison de la Radio", the restaurant Osmose forms part of a new generation of restaurant whose concept rests on the innovation of the menu and the decor. The fruit of a fusion of ideas between owner Franck Berrebi, chef Yoni Saada, and the architect and designer Olivier Gossart, the project gave birth to the first contemporary kosher, high end restaurant in the capital.

Thought of as a cocoon entirely moulded in plaster and covered with white and matt lacquer, the restaurant has a black parquet floor that contrasts and sets off the furniture created by Olivier Gossart. The bar in the form of a brilliant white pebble is in "osmosis" with the shell of the restaurant. Behind this, a curtain of water between the dining room and the kitchen allows one to make out the silhouettes of the kitchen staff.

The banquettes and chairs are lacquered white and covered in white fabric. The tables are in lacquered dark grey glass and initialled "Osmose". Lights are set into the plaster ceilings, which LEDs highlight the banquettes and create changing coloured ambiances. The below-ground floor is a world integrally treated with the colour black and a profusion of mirrors (walls, washbasins…). This design gives a very original spatial dimension. The only colour comes in the form of the words "Occupied" projected on the glass doors of the cubicles. Blue for the boys and pink for the girls!

毗邻巴黎广播大厦，欧思摩兹属于这新一代极为注重革新菜单与室内设计风格的餐厅之一。餐厅主人法兰克-贝若比和主厨尤尼-萨达的相识，加上建筑设计师奥利维耶-葛萨的参与，三人面对整体方案的态度与想法十分相投，于是巴黎第一家走高档路线、设计新颖前卫的犹太餐厅便诞生了。

设计师将餐厅构思成一个蚕茧般的空间，整体室内用纤维灰浆模制并覆盖一层白色雾面漆，而黑色镶木地板正好与其产生对比，并衬托出由奥利维耶-葛萨设计的家具。吧台设计成鹅卵石形状，与整个餐厅的蚕茧构思有异曲同工之妙，不过吧台采用的是白色亮面材质。吧台后面设置了一幕水帘，介于用餐区和厨房之间，客人可透过水帘瞥见厨师们的身影。

长形软垫沙发和椅子都漆成白色并采用白色罩套。餐桌是黑灰色漆玻璃并印有欧思摩兹的标记。灯光全都嵌置在纤维灰浆制的天花板里，一个个显示屏衬出长排软垫沙发并创造出变化多端的色彩气氛。地下室完全以黑色调处理，加上一面面的镜子(墙面、盥洗台等)，给整个区域一种十分奇特的空间感，只有带有颜色的"使用中"字眼被投射在玻璃制的厕所门上，男厕是蓝色，女厕是粉红色。

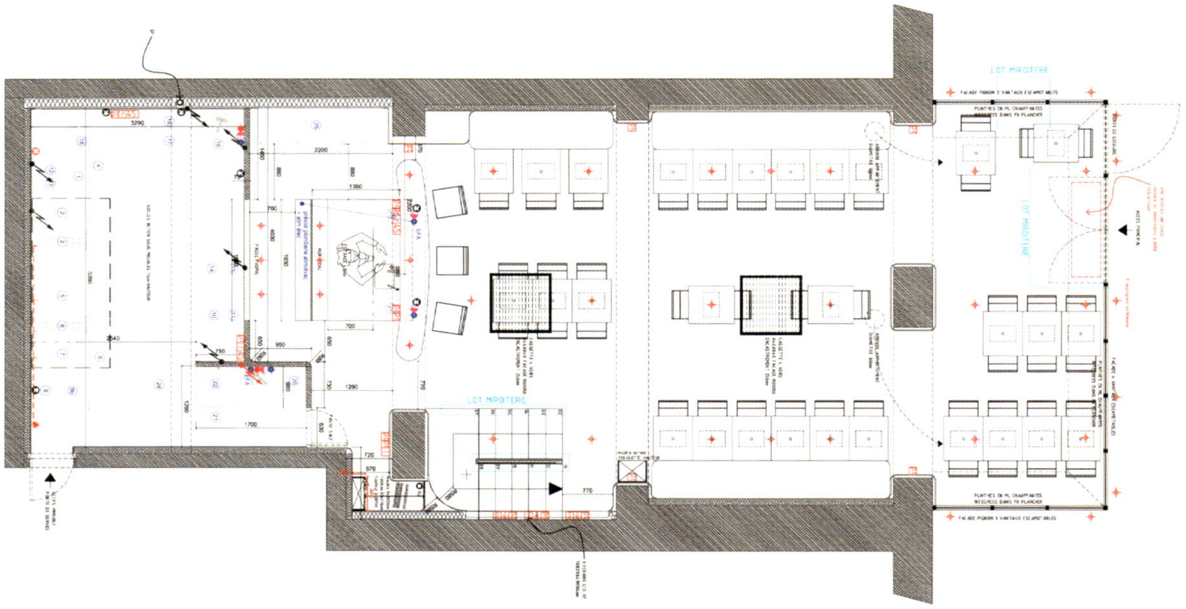

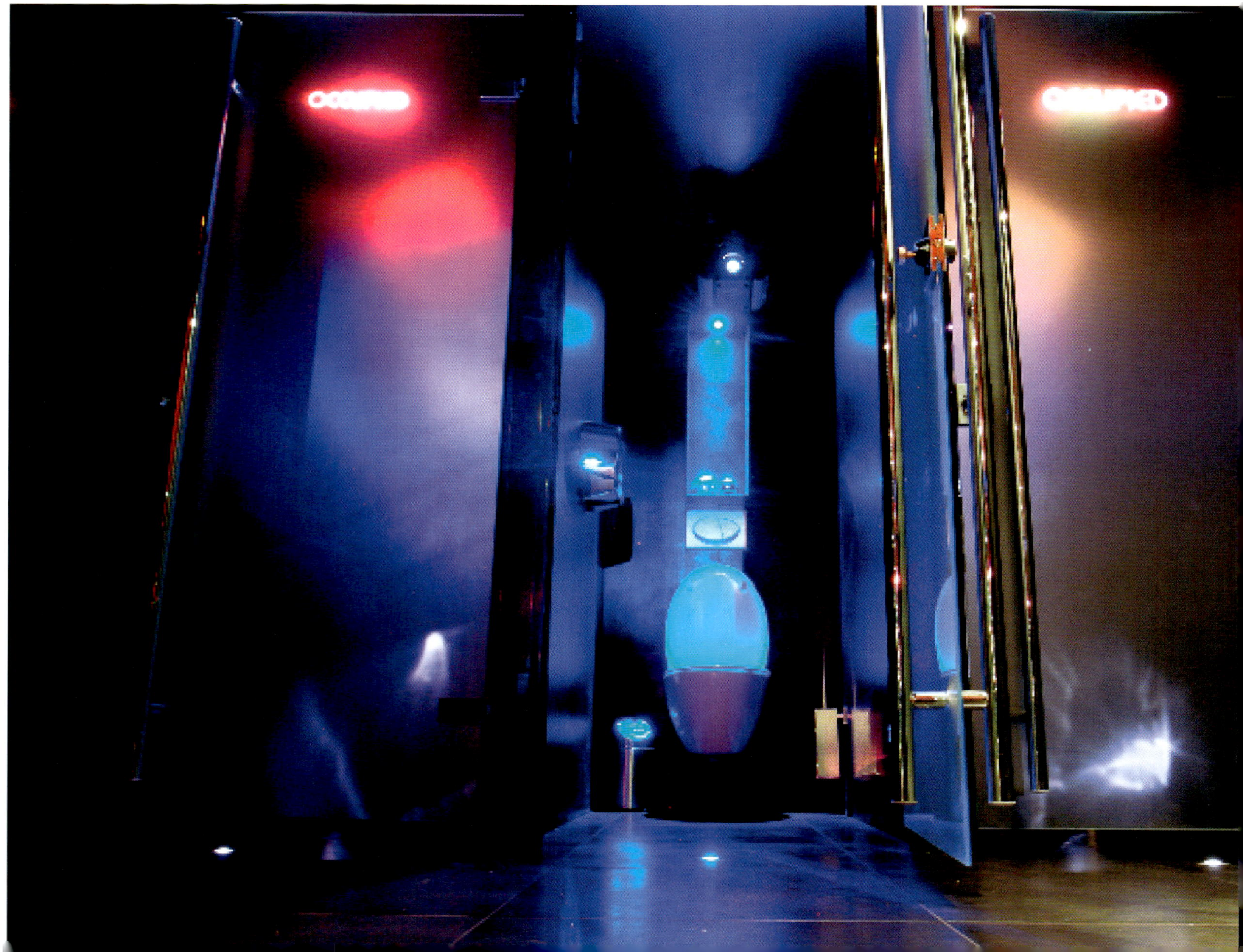
OCCUPIED

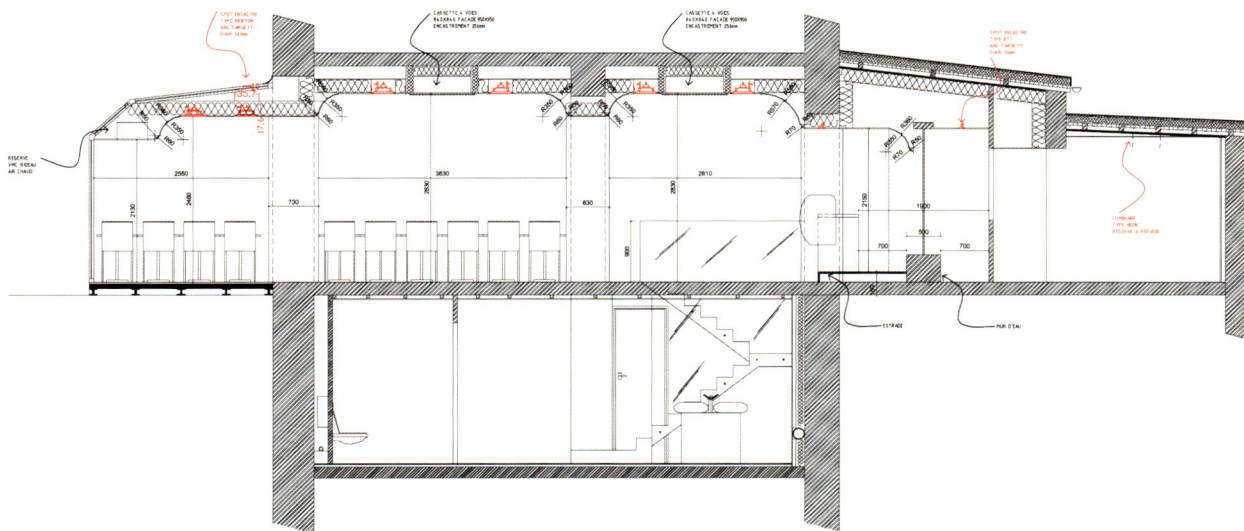

设计师 Designer : Patrick Jouin
地点 Location : New York, USA
完工日期 Completion Date : 2005
摄影师 Photographer : Eric Laignel

GILT

吉尔特餐厅

Gilt is an establishment with a feeling of light: from the very entrance, silvery pebbles evoke droplets of mercury. The structure that embraces the bar has an almost impalpable quality, like crystallization taking place; an unresolved tension that responds both to the domes of Buckminster Fuller and the stained glass windows of the patrician villa that stands behind it.

Patrick Jouin's mission was to install a new occupant in the venerable Villard House. A landscape with multiple references, the main space of Gilt is indisputably an historic monument, vouched for and preserved with an energy that tolerates no approximation. It is a rebellious environment: you can't drill, or cramp, or hang things willy-nilly, but at the same time a host of generous and eclectic fantasies have been summoned up: Greek or Roman antique styles, influences that come directly from the Renaissance and the Baroque, panelling and mouldings, marble and gold.

Patrick Jouin has obviously chosen the scenario of collision, but not of explosion. The installation must be precarious, ephemeral. Modernity, today, adapts to contradictory excesses and Patrick Jouin rejoices in and reverently respects an inherited plan. He will allow the Gilt to go beyond the stage of revamping, to go beyond a certain withdrawal. He makes it into a brilliant guest, one that participates actively in the conversation, without talking for the sake of it. Perhaps it is this that makes it seem so familiar straight away. It has never been there. It has never stopped being there.

吉尔特是一间设置在庞大雄伟环境中，但自身却轻盈现代的餐厅，其入口处的银色圆，令人联想到一颗颗的水银珠。厅内一切装修都具有不可捉摸的特质，环扣着吧台的结构体似乎正在结晶过程中，这未完成的张力却同时呼应着巴克明斯特-富勒建筑的圆顶和位在结构体后方的"贵族别庄"的彩绘玻璃。

设计师帕特里克-朱伊恩的任务是：在历史悠久的维拉德宫（纽约豪华酒店）里建构一个新空间。这个众人皆知的历史性建筑，是一个具有多重意义的城市地标，向来受到政府的积极保护，因此在其中进行的一切工程必然要求精准，不能允许任何的不确切。在此，建筑师不能随意凿穿墙壁、锚固或吊挂，装修极为受限；但同时，整体环境又是如此地充满奇特的想象力，大方又兼容并蓄，有着古希腊和罗马风格，也有直接受文艺复兴及巴洛克影响的风格。处处是古色古香的护壁镶板和线脚装饰、高贵的大理石和金碧辉煌的装饰。

帕特里克-朱伊恩显然选择了运用古今之间的对立来产生对比美感，但并非过度冲突所产生的爆发性场面，因此，此餐厅内的装置必须是短暂的、具有昙花一现般的美感。今日的现代艺术充斥过多的矛盾，但帕特里克-朱伊恩却乐取其菁华，在尊重传统格局的前提下发挥现代设计精神，因此使得吉尔特餐厅的室内重新整修能够超越一种只是除尘埃或诚守缄默的做法。他使得餐厅在这古老的环境中像个出色的宾客，积极参与对话，却又不至于过度饶舌。或许正因此，此餐饮空间能够令人在即刻间产生亲切感。它似乎从不存在，又似乎从未停止存在。

设计师 Designer : Patrick Jouin
地点 Location : Illhaeusern, France
完工日期 Completion Date : 2007
摄影师 Photographer : Eric Laignel

L'AUBERGE DE L'ILL
伊尔河客栈

Located on the banks of the Ill, the river that crosses Alsace, L'Auberge de l'Ill has been welcoming food lovers since 1879. On his arrival, the designer Patrick Jouin immediately noticed that the cuisine and the atmosphere were directly linked to the local history and quality of the land.

The ebb and flow of the clientele in the establishment was the first thing to be improved. The glass-roofed entrance porch, clad in horizontal wooden laths like the traditional local tobacco drying sheds, draws attention to the attractive main door. A pathway of ceramic tiles and brushed carpet leads visually towards a wrought-iron sculpture of a fish. Inside the main lobby visitors are greeted by a reception desk covered by a felt mesh whose main motif is the new logo, and a deep red salon with furniture designed by Patrick Jouin.

In the two first rooms, "the veranda" and "the dining room", the hand-tufted carpet represents a symbol aerial view of Alsace: different tones of dark green on closely layered contours give one the impression of being at the heart of the Vosgien forest. These two rooms are lit by glass tubes planted in pools of mirror-polished stainless steel, as if the frequent floods had left puddles here. The padded wall covering stops short of the door frames, which are without curtains, freeing up the view of the gardens and the river.

The rooms that follow evoke peasant architecture: the floor coverings change and raw wood is far more present. In the "Alsatian room" an entire wall has been covered with large planks of brushed pine and paintings by Roger Muhl. In the "pigeon loft" the walls and the sloping ceiling have been covered in the same pine planks, and the entrance is endowed with a large double sliding door garnished with branches of chestnut and hazel and giving the atmosphere of a barn…

Serious thought has also been given to the lighting, whose intensity and colour evolve as the meal progresses. In the second half of the evening, the intensity slowly resides towards a very warm atmosphere, while, in counterpoint, LEDs change from amber to light blue.

伊尔河客栈坐落在流经阿尔萨斯地区的伊尔河河畔。自1879年以来，客栈接待了不少前来的游客及老饕。设计师帕特里克-朱伊恩很快就理解到此地餐厅的气氛和美食跟当地的乡土与历史有密不可分的关系。

客栈重新装修后，吸引了络绎不绝的宾客。餐厅入口处用玻璃架构出来的玄关空间外围装了水平向排列的木条，就像当地的老式烟草干燥屋的外观，使客栈的大门更为醒目。入门后，狭长的砂岩地板和一条擦鞋垫毯将视线引至一尊鱼形雕塑，是手工制的铁饰品。经过玄关，一座接待柜台铺了毛毡制的格栅，上头的基本图案便是客栈的新商标。紧接着，是一间深红色的接待厅房，摆置了帕特里克-朱伊恩为卡西纳设计的家具。

接下来位于前面的两间大厅房是"游廊"与"饭厅"，手工制地毯的图案呈现出阿尔萨斯地区的鸟瞰概貌，运用不同色调的深绿配上紧密的金黄色等高线，让人感觉恍如身置孚日森林中。这两间厅房的灯光来自一根根芦竹般的玻璃管，种在池沼般光滑镜面的不锈金属中，犹如频繁的涨水留下了一坑坑的水洼。墙面所覆的软垫一直铺到贴齐窗框为止，没有装窗帘，留下窗户的大片玻璃空间，将窗外花园及河流的美景尽收眼底。

再进去的几间厅房的风格很接近农村的建筑，地板的材质与前面的厅房不同，运用较多的原木材料。在"阿尔萨斯屋"中，一整面墙都铺满了刷饰过的大片冷杉木板及罗杰-穆勒的图画。在名为"鸽棚"的厅中，墙面及斜柱也都覆盖了同样的冷杉木板，入口还装置了一个大滑动拉门，两扇横栅式门扉饰了栗树及榛树树枝结构，营造出犹如谷仓的气氛。

整个客栈的灯光设置也作了深入的改装，跟随用餐进程，灯光的强弱与色彩也跟着改变，到了晚餐的后半时段，灯光渐趋柔和，营造出温馨的氛围，而相对地，LED灯也渐渐从琥珀色转为淡蓝色。

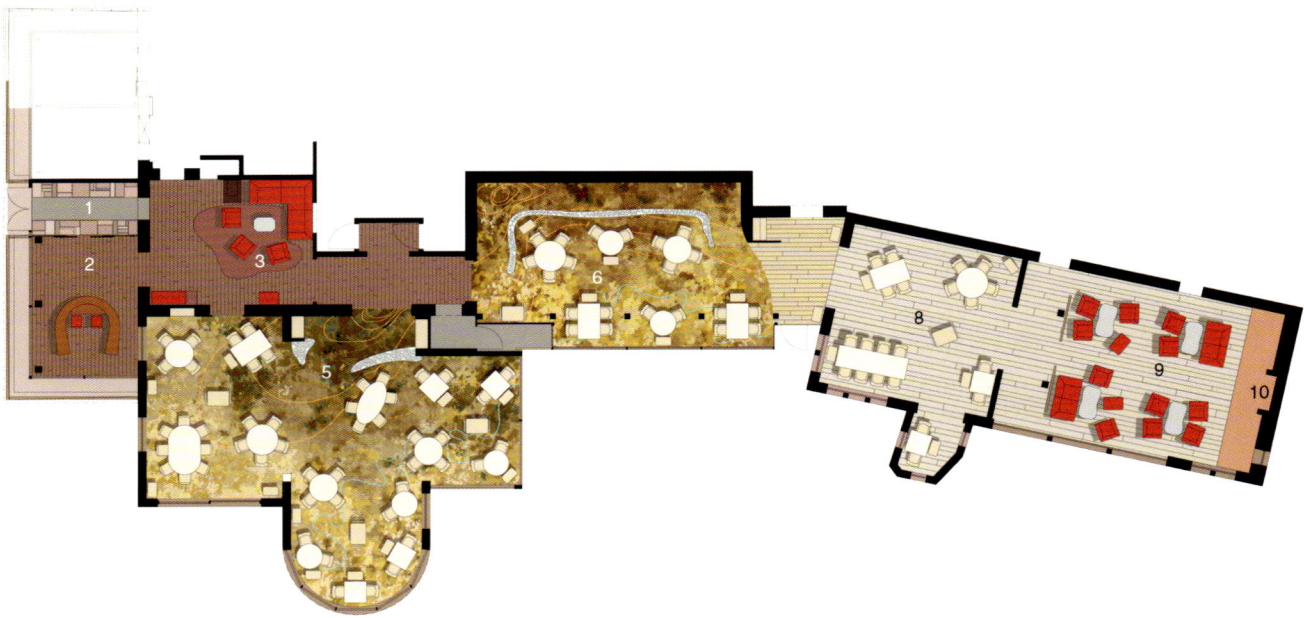

设计师 Designer : Philippe Starck
地点 Location : Paris, France
完工日期 Completion Date : 2008
摄影师 Photographer : Francis Amiand

MAMA SHELTER

妈祖庇护酒吧餐厅

The authentic Paris of the Saint Blaise neighbourhood, with its café terraces, its gardens and its rebellious atmosphere, was chosen as the location for Mama Shelter, a "living place" with a unique atmosphere. Dreamed up by the Trigano family (co-founders of Club Med) and the philosopher Cyril Aouizerate, Mama Shelter was designed by Philippe Starck.

More than a place to sleep, Mama Shelter, an urban kibbutz or laic monastery, is a place to get together over a shared dish. On the ground floor, a gigantic communal table inset with video screen welcomes travellers. Even if they are alone, visitors find themselves beside or face to face with an "other". In front of this table, refrigerated storage units dish up "Sweet or Savory Dinner" 24/7 with a corner devoted to confectionary. The guests can serve themselves or settle into the armchairs near the fireplace, in front of the candles that never burn down. In the middle of the tables, columns house screens where you can place messages to chat, meet people, or simply dream.

This restaurant is accompanied by a giant double bar, the "Chic-Chic Bar", a brasserie and private terraces where American poets, Japanese painters or Latin American writers cross paths.

"妈祖庇护"酒店及其酒吧餐厅位于圣布雷兹区，是一个真正很巴黎的城区，有一家又一家的露天咖啡和一座座花园，且处处弥漫着叛逆的气氛。"妈祖庇护"是个风格独一无二的生活空间，由崔伽诺家族（地中海俱乐部的创始者之一）和哲学家西瑞尔-阿维兹瑞特共同创立，并请名设计师飞利浦-斯塔克负责装修设计。

"妈祖庇护"不仅是个可下榻住宿的酒店，更像是城市里的基布兹农村或非宗教的寺院，是一个可以相聚一堂共享一餐的好地方。在地面一楼有一张桌面镶嵌了录像荧幕的特特大号主宾桌，迎接前来的客人。在此，即使有人独身而来，一旦坐在这张大桌旁，就也一定能够面对着"另一个人"。在这张长桌前方，排列一个个冷藏柜，极像那种24小时全日开放的"自助餐饮店"，还附带一个甜食区。客人可自行取用柜中的餐点，或坐在靠壁炉的沙发中，倚在烧不尽的烛火前。在一张张餐桌之间，几根柱子上嵌置了荧幕。客人可以在此传放广告做交流或交友，或只是单纯做做梦。

此餐厅附设有一间大型酒吧，酒吧内还有一个"摩登摩登吧"、一间酒馆和一个贵宾露天座。在此酒吧，人们常会遇见一些美国诗人、日本画家或南美作家。

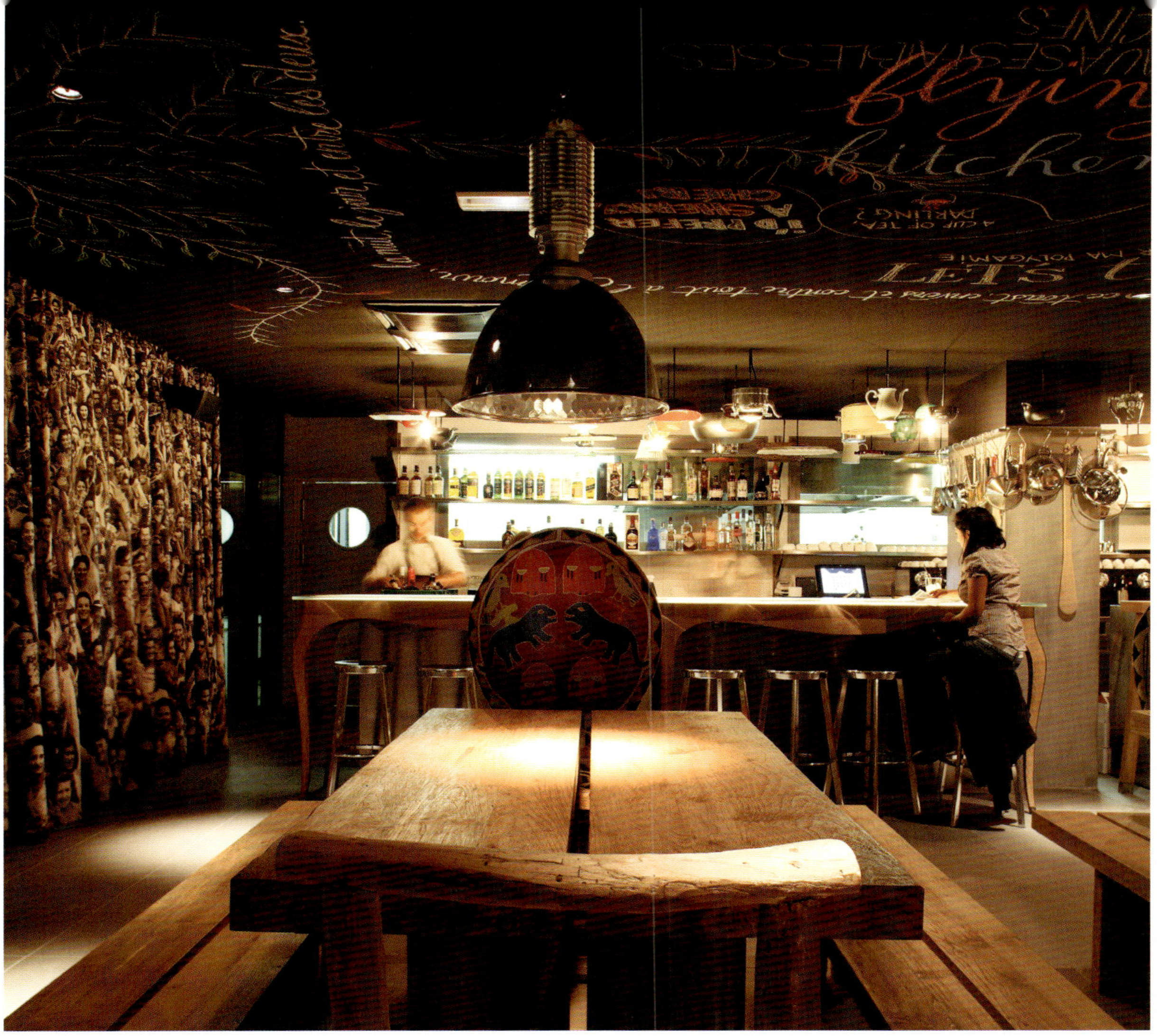
kitchen

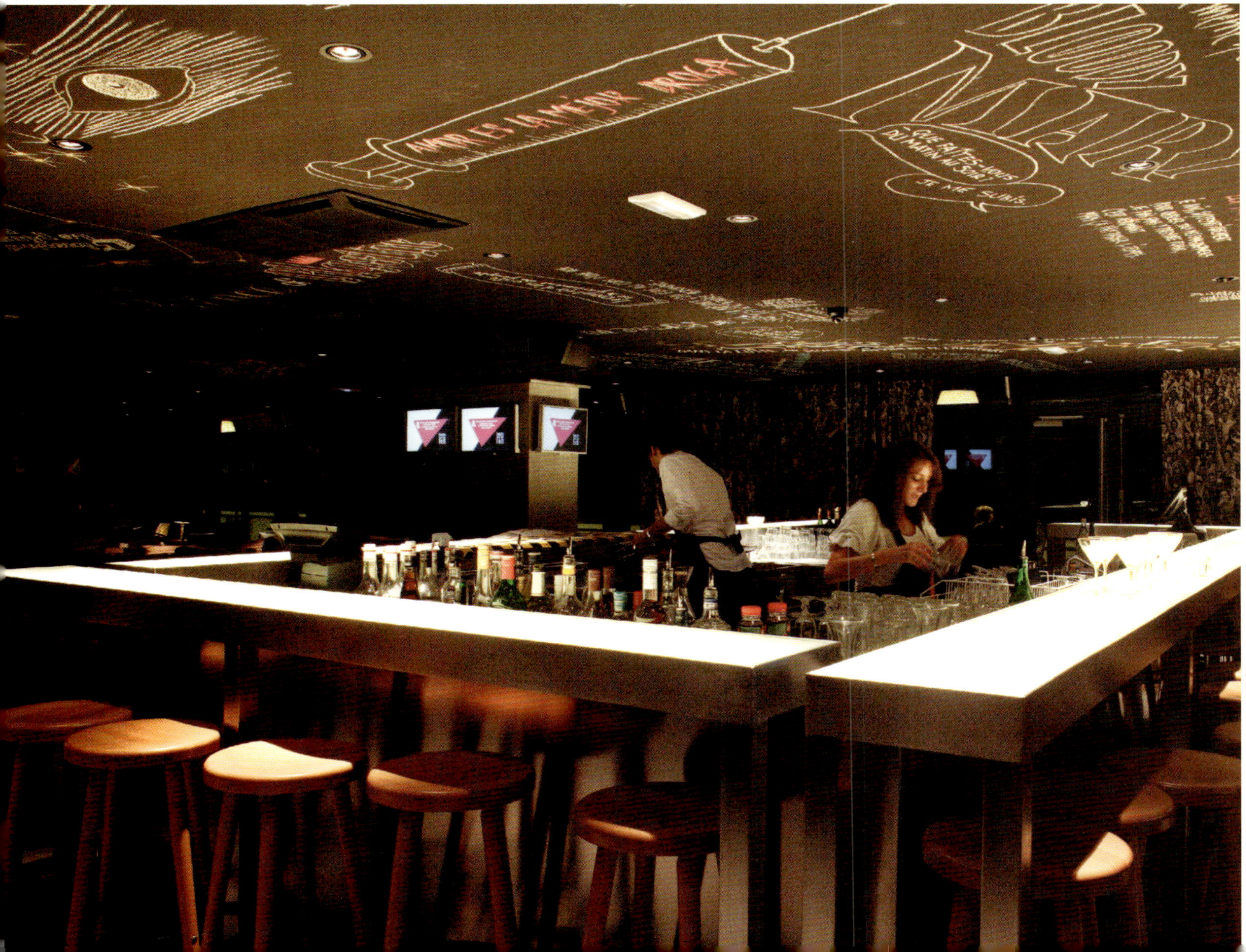

设计师 Designer : Valérie Serin
地点 Location : Paris, France
完工日期 Completion Date : 2008
摄影师 Photographer : Yvan Moreau

NABULIONE

拿破里昂餐厅

The restaurant Nabulione has an extraordinary setting facing the Dôme des Invalides, at the heart of one of the most confidential neighbourhoods of the French capital. The guest of honour is Napoléon, who is called Nabulione in Corsican. The story recounted here is like a private visit to the Invalides, in particular its Dôme and Salon d'Honneur. The place gives the impression of having always been there, and only lightly touched by the designer.

The presence of Napoléon is invoked by porphyry, the marble of Emperors. All the architectural and decorative codes suggest the Dôme (the pilasters, glass ceiling, painted ceilings, marble floors and tombs), and the Grand Salon d'Honneur, with its chimneypiece and panelling. These details reinforce the intimacy of the place, in particular the mounded doors that give the impression that the restaurant continues. On the ceiling, fabric is stretched giving the illusion that frescoes have been painted there. They depict details from the 18th century Apotheosis of Hercules from the Hercules Salon of the Chapel of Versailles, and the 17th century Glory of Saint Ignacio Loyola in Rome.

The deliberate choice of contemporary furniture and lighting brings a surprising and comfortable character to this classical setting, which is both luxurious and glamorous. The two covered terraces mark a transition to the outside terrace, which is spacious and tree-filled. It is inspired by the lines, rhythm and hedges of the Invalides gardens. The planting containers also prolong this inspiration.

拿破里昂餐厅有着格外特别的地理位置：面对着巴黎荣军院的金色圆顶，位处法国首都机密性最高的地区中心。餐厅的荣誉贵宾、其礼赞的对象就是法国历史英雄拿破仑，其家乡柯西嘉方言则发音成拿破里昂。宾客来到此餐厅就犹如私下参访荣军院，尤其就像参观了圆穹和院中的荣誉厅。这地方让人深深感觉到它似乎自古已存在，只有轻微的整修过。

在此处处让人想起拿破仑，尤其是斑岩材质这个代表皇帝的大理石。厅内整体的建筑空间及装饰特色都令人联想到荣军院里的圆穹（壁柱、玻璃天棚、壁画天花板、大理石地面、陵墓），还有像在荣誉厅里看到的壁炉和雕木装饰，强调了餐厅的温馨亲密氛围。门上精致的线脚装饰延伸了空间感，让人觉得门后好像还有另一个餐厅空间。天花板上的画布紧绷裱在上面，让人误以为是直接画在天花板上的壁画。画里某些细节描绘的灵感来自凡尔赛宫中的18世纪的画作：海格里斯厅的壁画"海格里斯的赞颂"以及和皇家礼拜堂的壁画；还有来自在罗马的17世纪作品"圣依纳爵罗耀拉的荣耀"。此装修工程都交由阿里欧斯公司来完成。而照明气氛的营造则要归功于克里斯提安-布若吉尼。

使用极具时尚风格的家具与灯饰的做法，为这个风格古典的餐厅带来既舒适、豪华、高贵却又有点叛逆味道的魅力。两个有遮顶的露台座成为室外那宽阔且绿意盎然的露天座的过度空间。室外露天座的设计灵感来自荣军院花园的节奏、线条和黄杨的栽种。这里选用植物形制的家具也延伸了来自花园的灵感。

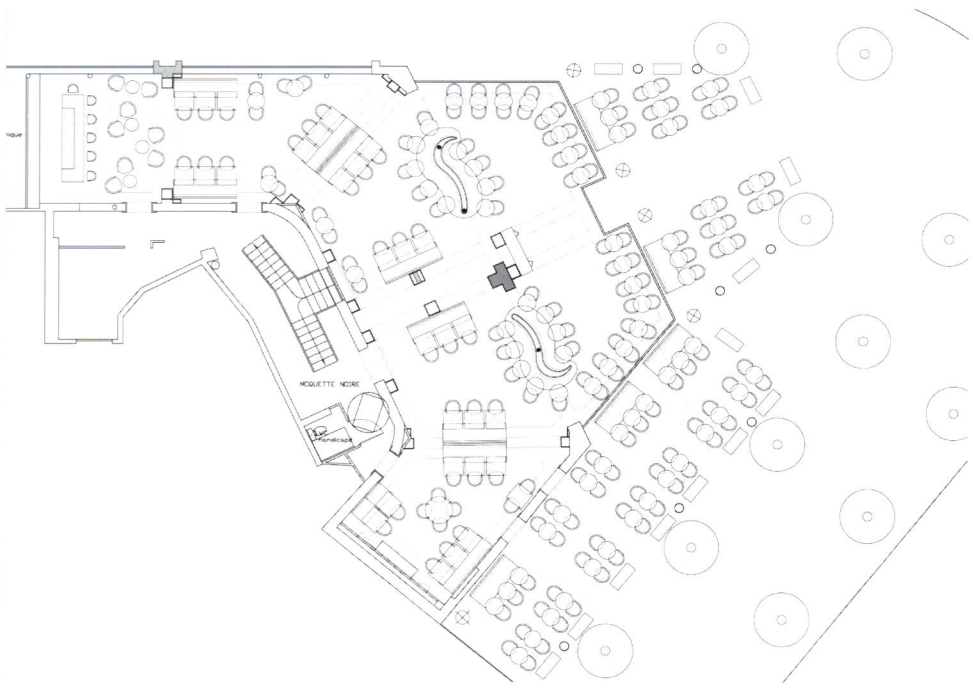

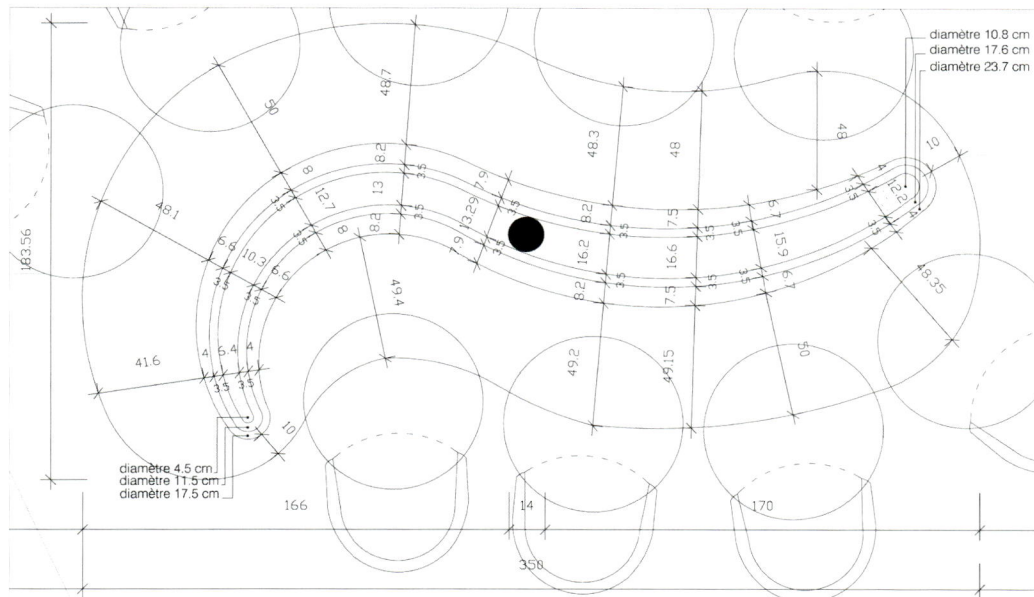

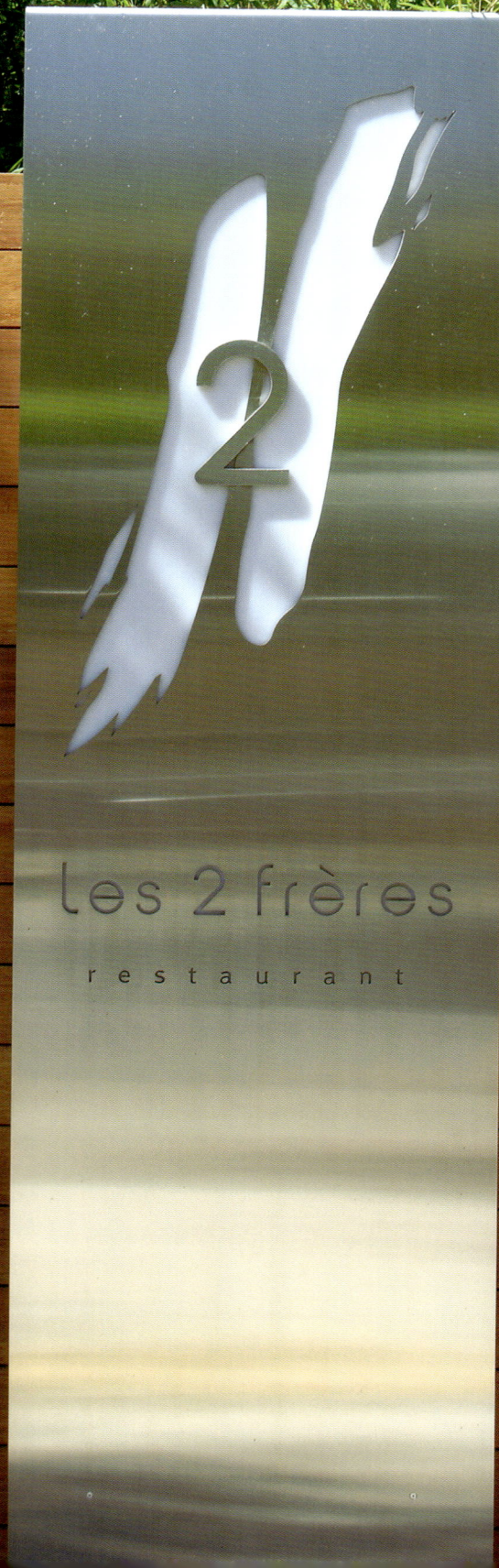

les 2 frères
restaurant

设计师 Designer : Xavier Luvison &
Jean-Christophe Sabarthès
地点 Location : Aix-en-Provence, France
完工日期 Completion Date : 2006
摄影师 Photographer : Les 2 Frères,
Agence Caméléon

LES 2 FRERES

兄弟餐厅

The exteriors of the restaurant Les 2 Frères are defined by high boarding in exotic wood and by bamboo hedges lit up in an indirect play of light. Jean-Christophe Sabarthes is the architect: "My project leans on the properties of simple, geometric forms, and the idea was to create a spare but welcoming decorative scheme. For that I have used common elements in the interior and exterior decoration and "real" materials such as glass, wood, metal and concrete."

The main entrance of the restaurant is a portico in stainless steel clasped on either side by windows, responding to the serving hatch that is found opposite in a perfect symmetrical alignment. Above this entrance, an element added to the ceiling in the form of a "u" provides lighting for the whole room. The furniture, composed of rectangular or round tables, has the look of exotic wood. The light coloured exterior chairs are "Toys" by Philippe Starck. The round tables, prolonging the line of the main entrance, trace an axis with the wide serving hatch that provides a view into the kitchen area.

An elevated lounge occupies one side of the restaurant. It starts on the terrace, where it is shaded by fabric covered pergolas, and extends to the bar. Music and vegetation form an integral part of this privileged space. Two television screens embedded in a mesh of stainless steel show clients the dishes being prepared in real time in the kitchen. At the centre of the dining room is a third, giant screen; two pieces of glass going from floor to ceiling clasp the translucent screen, which receives the same image, a key decorative and informative element.

兄弟餐厅的外围是运用带异国风味的进口木材所制之高墙和一丛丛竹篱来界定空间，并搭配了间接照明系统所营造出的灯光效果。负责此设计工程的建筑师让克里斯多夫-萨巴特诠释道："我对此方案的构想着重在呈现单纯的几何形体，意在创造简洁优雅却不失温馨的装修空间。为此，在室内及室外，我都使用相同的建材，运用玻璃、木材、金属和混凝土这些真材实料来构建。"

餐厅主要入口处设计成一道不锈钢与玻璃组成的门廊，而门廊正对面就是餐厅与厨房间的递菜窗口，门廊与窗口两者间构成一条完美的对称线。在这个玄关上方的天花板上加了一块"U"字形的饰板来装设灯光，以照明整个厅内空间。长方桌和圆桌组成的家具的木材质感看起来很有异国风。放在室外并题名为"玩具"系列的浅色座椅是名设计师飞利浦-史塔克的作品。一排圆桌从玄关到递菜窗口列成一直线，而这窗口也故意设计得十分敞开，让客人可窥见厨房内部空间。

基底架高的沙发餐饮区占了餐厅一边的空间，从布棚下的露台开始，一直延伸到吧台。背景音乐和花草植物也扮演了这个舒适的餐饮空间中不可缺少的重要角色。顾客可以从两座嵌置在不锈钢网上的电视荧幕上看到厨房里料理美食的同步现况播放。在餐厅内的中心位置上还有第三座超大荧幕，从地上到天花板，由两片玻璃夹着一面半透明的屏幕组成，荧幕上也播放同样的料理影象。这些令人看了垂延三尺的画面，不但成为餐厅内引人注目的装饰，也提供给宾客美食料理的资讯。

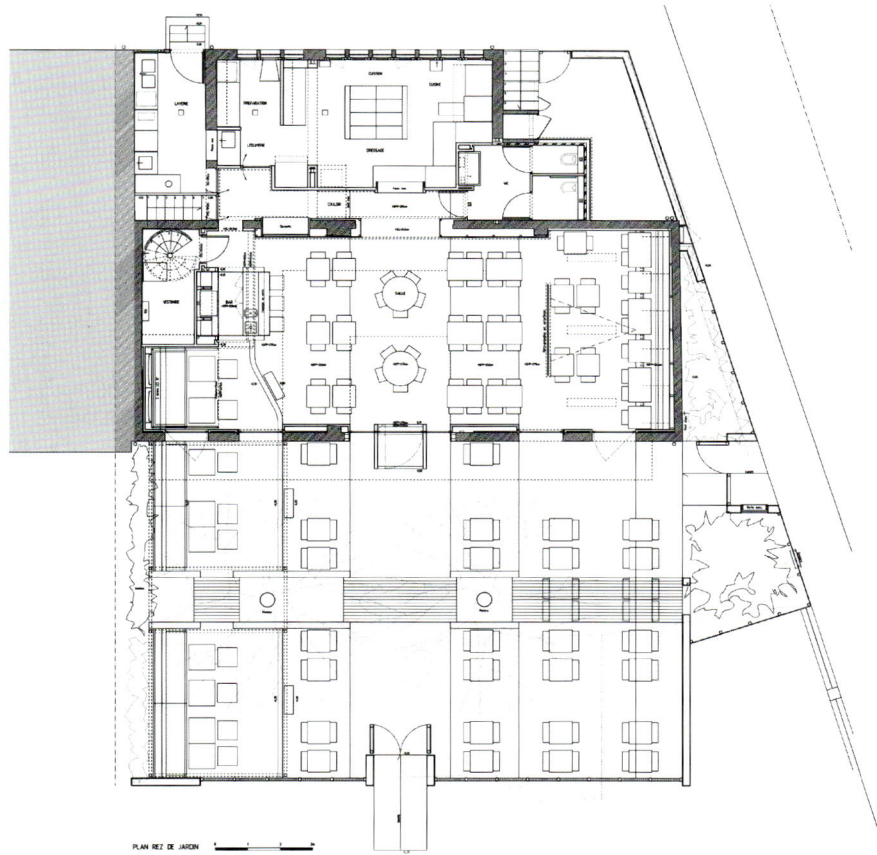

PLAN REZ DE JARDIN

设计师 Designer : Christian Liaigre
地点 Location : Paris, France
完工日期 Completion Date : 2001
摄影师 Photographer : Vincent Fillon

MARKET

玛克特餐厅

Christian Liaigre had already worked very closely with Jean-Georges Vongerichten, the owner of Market, for the decor of Mercer Kitchen in New York and of Dune in the Bahamas. For Market the original idea was to emphasise the notion of soil, foster mother and origin of life. He therefore suggested the use of raw materials such as wood and stone. Very few objects are on view, only some beautiful masks and shields from the tribal art of Africa and the South Pacific.

The architect prioritised the colour grey for the panelling and stone elements, and off-white and transparency for the crockery. All these things are highlighted to advantage by precision lighting. A bar welcomes guests in the hall and a "raw bar" is used to present oysters, fish carpaccios and other raw preparations.

The crockery is made of off-white stoneware, with transparent modern glasses. There are no flowers on the tables but candles reinforce the warm feeling the place inspires. The Mercer Kitchen has a special feature: one can see the kitchen from the basement where the wine cellars and the "chef's table" are situated.

建筑师克里斯提安-里耶格和巴黎玛克特餐厅的老板让乔治-佛洁瑞西顿已经有过合作经验，以绝佳的默契创造了纽约的媚瑟膳房以及巴哈马的沙丘餐厅。玛克特餐厅的原始构想是要呈现"土地"的概念：土地是孕育大地之母，也是生命的起源。因此，设计师建议运用较简朴的材质，例如木材和石材，来作为空间装修的主要材料。餐厅内饰品不多，只有几面来自非洲或大洋洲的精美原住民面具以及原始艺术盾牌。

建筑师特别运用灰色系来强调空间木质和石材的装饰，而餐具则采用乳白色以及透明的材质来配合点缀，整体在精确设计的灯光搭配下展现出了最佳效果。靠入口空间旁有个吧台，迎接入门的客人，另一个生鲜吧则摆设了生蚝、生鱼片和其他的生鲜美馔。

乳白色粗陶制的餐具搭配上极具现代感的透明酒杯，桌上不放任何花饰，只有光影摇曳的蜡烛，增强此餐饮空间的温暖气氛。这里有类似媚瑟膳房的空间处理：在通往地下室酒窖和名厨宴客桌的路程中，可以一窥厨房的堂奥。

设计师 Designer : Atelier FB
地点 Location : Paris, France
完工日期 Completion Date : 2007
摄影师 Photographer : Arnaud Rinuccini

HANAWA

汉纳瓦餐厅

The challenge was to transform a former showroom a few steps from the Champs-Elysées into a prestigious Japanese restaurant seating almost 200. After a year of planning and 18 months of work, food lovers now come to taste a contemporary Japanese cuisine mixing different culinary traditions in a discreetly luxurious setting.

The premises (1500m² spread over three floors) is composed of nine spaces with distinct themes, each zone being fitted out specifically in line with the cuisine on offer. A vast ground floor welcomes visitors and houses a tea salon with a serene ambiance in counterpoint to the delicate pâtisseries. At the three below-ground counters with their futuristic design, chefs prepare a gastronomic Western-style cuisine cooked on heated plates in front of the clients – Teppanyaki.

On the first floor, three rooms are arranged around a Japanese Tsubo-niwa garden (an interior courtyard), where lanterns in lava stone and miniature vegetation are in harmony with the traditional Japanese cuisine. The sushi bar with a sea theme is open to the two communal tables that each seat 12 guests, where the mouthfuls are served freshly chopped. A choice of fine materials such as the bronze of the entrance partition, the okoumé wood of the parquet or the untreated Japanese elm on the walls, gives the place a sober elegance inspired by Zen.

这是一个十分具有挑战性的室内设计方案，经过一年的研究，加上18个月的施工，将这间毗邻香榭丽舍大道的旧展示厅转变成一间可容纳近200位宾客的高级日式餐厅。赏识美食的老饕纷纷前来，在含蓄奢华的环境中品尝最时尚的日本餐，体验日本各种不同传统的厨艺。

餐厅有三层楼，共1500平方米，由九个不同主题的空间组成，每个区域都依照其所提供的餐点而作出特殊的空间设计。迎接宾客的是一个广大的地面层空间，这里设有一间气氛宁静的品茶沙龙，搭配着精致多样的糕点。地下楼层的装修具有未来主义设计风格，厅内有三座铁板烧台，师傅当着客人的面，在铁板上烹烧着西式的铁板烧美食。

位于二楼的三个厅房围绕着一座日式内庭花园，园内熔岩石制的塔灯和盆栽植物与餐厅所提供的日本传统美食和谐呼应。弥漫着海洋气息的寿司吧旁摆设了两张大主宾桌，各有12个位子，客人在此可享受现做的新鲜寿司。餐厅装修选用了多种极为高贵的材质，入口玄关的隔板是铜材，地面是奥克曼木材，墙面采用日本的榆树原木，营造出此餐厅充满禅境的高雅气质。

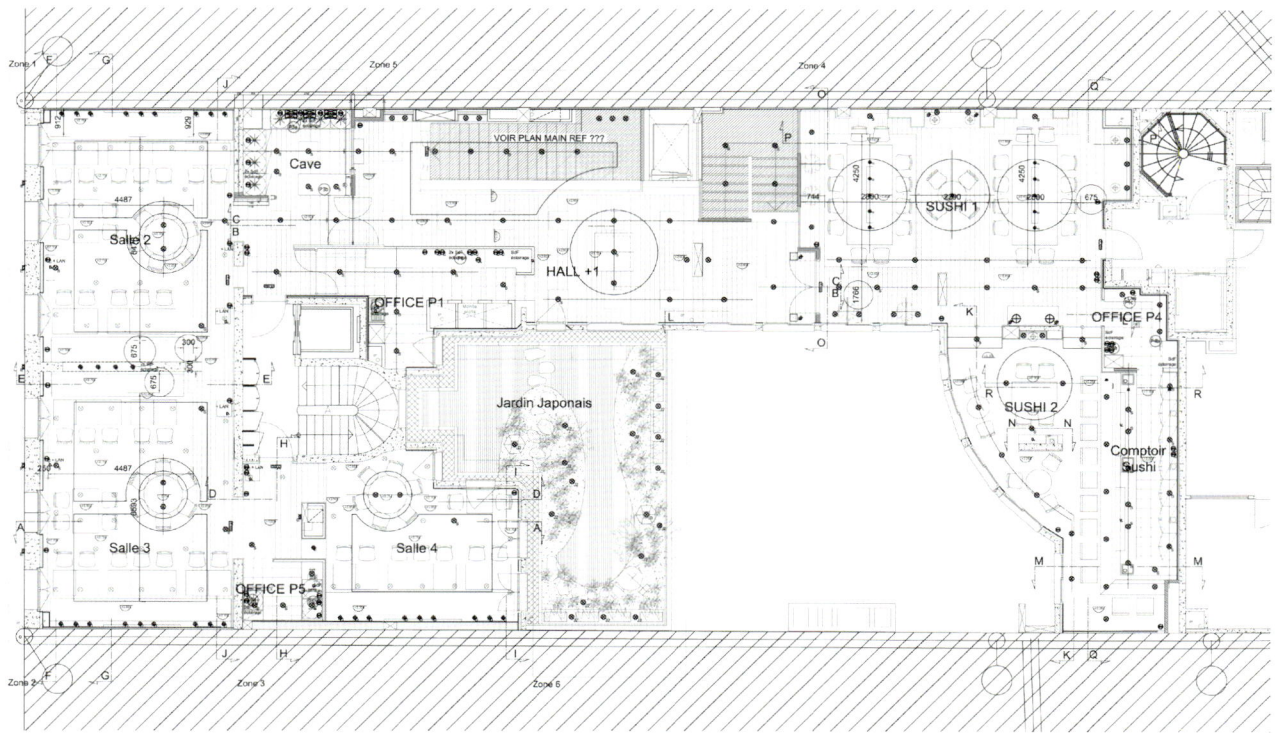

设计师 Designer : Pierre-Yves Rochon
地点 Location : Tokyo, Japan
完工日期 Completion Date : 2005
摄影师 Photographer : Kayoko Nonamura,
Tetsuya Miura

BENOIT

波诺瓦餐厅

On the 10th floor of a Tokyo building, Café Benoit begins with an impressive French-style bar three metres long at its entrance. Created by a French master joiner, it is inspired by the bars of early 20th century, with its pewter counter and walnut casing. The decoration as a whole rests on a choice of quality, authentic materials: oak Versailles parquet, cement tiles with traditional motifs, 18th-century terracotta, oak panelling, walnut bistro tables for the upper floor and pewter ones for the bar, shelves stocked with old jars, glazed terracotta pots and country recipe books...

The 11th floor offers an astonishing view over the city through a large window, while its interior evokes a French bourgeois residence. The visitor immediately feels at home on seeing the desk, the private dining room, the collection of objects and paintings, the distressed stone flooring, the old-fashioned patterns of the fabric lining the walls... At the back of the room, the winter garden is an invitation to let time stand still, and the large wrought-iron clock has stopped.

In the private dining room, the "charcuterie ceiling" of painted fabric dates from the beginning of the 20th century. Here and there unusual objects are on display: a metre-high model of the Eiffel Tower made by Gustave Eiffel, bronze animal sculptures by Nicole Cane, photographs by Jean-Louis Bloch-Lainé taken from the book La Riviera by Alain Ducasse...

波诺瓦咖啡厅位在一座大楼的10楼，客人一进门会先看见一座令人印象深刻的法式酒吧，吧台长达3米。此吧台是由一位法国高级木器师傅精心制造而成，灵感来自20世纪初的吧台风格，锡制台面搭配着胡桃木制的吧台柜。厅内整体装修皆选用上等的真材实料，如：橡木制的凡尔赛式镶木地板、饰古典图案的水泥砖、18世纪的陶土、橡木制的细木护壁板、楼上餐厅配置胡桃木制的餐桌，酒吧区则采用锡制桌椅，一层层的柜子上摆饰了各种古老的瓶瓶罐罐、上釉的陶壶和谈论具地方特色的美食书籍。

位于11楼的餐厅空间拥有一座大玻璃天棚，客人由此可欣赏到令人惊异的城市美景。厅内装饰令人联想到极其法式的布尔乔亚家居风格。客人一来到此厅，会感觉犹如回到自己家中，看见书桌、私人饭厅、收藏品和一幅幅画作，这里还有铺陈了古旧石板的地面，以及图案古色古香的墙饰。在餐厅最里面有间玻璃花房，盛邀客人来此休憩放松，连厅内的那一座锻铁大时钟都已停摆。

私人饭厅里挂着一幅上世纪初的裱画作品"猪肉食品天堂"，呼应着这里到处展示的稀奇古怪的物品，例如：一座由古斯塔夫-埃菲尔制作、1米高的埃菲尔铁塔模型，妮克-坎恩创作的铜质动物雕塑，还有出自"阿朗-杜卡斯的蔚蓝海岸"一书中，由让路易-伯罗旭雷内拍摄的摄影作品。

设计师 Designer : Code Déco
地点 Location : Paris, France
完工日期 Completion Date : 2008
摄影师 Photographer : Gilles Dacquin

LE TELEGRAPHE

电报餐厅

Classed as a Historic Monument for its architecture and its Art Nouveau stained glass, the Télégraphe has reopened. Its superb facade and its inscription "Maison des Dames des Postes and des Télécommunications" ("house of the women working in the post and telecommunications") have to be admired. Two black and white lights adorn the entrance. In addition, a 4.8-metre-high ceiling, a glass roof, an interior garden and a cosy bar make it a unique place.

The owners and designer Denise Omer wanted to create a more contemporary setting. In the main dining room, the floor in marble mosaic and the wood panelling are listed. The wood was darkened to give a wenge wood colour. The pillars have a nickel tinted patina with multiple reflections, and the ceilings have been "lightened" with celadon green. A large piece of original wooden furniture "dresses" one side of the room. The chairs have a silvered leather effect. The wall lights with their sinuous forms diffuse a pleasant indirect light.

At the bar one finds a parquet that is also tinted to look like wenge wood. A turquoise and silver wallpaper decorates the walls, where two giant hanging lights are reflected on each side of the bar. Very comfortable club chairs in a nickel coloured vinyl give a more intimate feeling.

The walls of the conservatory are painted raspberry, and marry very well with the woodwork, offering dynamism and gaiety. Three Murano chandeliers "fall" from the glass roof that is fitted with a canopies. The wall lights in Murano glass and Swarovski crystal give a precious look. The chairs are in white pearlized vinyl. The view of the garden and the excellent light give a "Riviera" feel to the place.

The terrace is exceptional: a haven of peace surrounded by a Japanese garden. A pontoon, teak tables, seating in imitation woven leather and superb parasols decorate this little paradise. A Japanese fountain adds an extra Zen touch.

电报餐厅因为其建筑特色以及具有新艺术风格的彩绘玻璃，而被列为国家古迹级保护建筑，此餐厅近日才重新开幕。来到此地，一定要仰头欣赏这宏伟壮丽的建筑门面和标注"邮政电信贵妇之家"的铭文字号。两组黑白灯饰照亮大门，接着有天花板高度达4.8米的餐厅空间，加上大玻璃廊厅、室内花园和温馨的酒吧，创造出一个独一无二的餐饮空间。

餐厅的股东们偕同设计师德尼斯-欧美，希望创造一个更具现代感的时尚空间。在主厅房中，大理石马赛克地板和细木护壁板都列入国家保护名单，因此，细木只被加深色调，呈现类似红木的色泽。所有支柱都带有镍色泽与多重反光的质感，天花板则被处理成淡绿色以提高明亮度。一件原本就存在此地的大型古木家具成为厅内部分空间的装饰物。这里的座椅呈现一种银色皮革的质感，而形状弯弯曲曲的壁灯则散发出非常柔和的间接照射光线。

酒吧区的镶木地板也是沉稳的红木色。墙壁贴了青绿配银色的壁纸，酒吧两边各饰了一盏巨大的吊灯，灯光悠然地映在壁纸上。镍色乙烯基（俗称维尼龙）制的舒适沙发椅为酒吧带来一种更亲密的氛围。

大玻璃廊厅的墙柱漆成覆盆子色，与细木护壁板的感觉非常搭配，也为空间增添了活泼喜乐的感觉。三盏华丽的穆哈诺水晶吊灯从天花板垂吊下来，而以穆哈诺玻璃和施华洛世奇水晶合制而成的壁灯也赋予了餐厅尊荣华贵的气质。此厅的扶手椅由带有珍珠光泽的白色维尼龙材质制成，为空间添增雅致柔和的氛围。厅内可欣赏花园的美景，加上耀眼的自然光线，令人仿佛置身于一家蔚蓝海岸的餐厅。

餐厅还有个非常难得且独特的露天座，像个和平的避风港，四周围绕着极具日本风味的花园。这里有浮桥、柚木餐桌、仿皮编织座椅，再缀上漂亮的阳伞，营造出一座迷人的乐园；一座日式喷泉更为此地增添了禅的意境。

设计师 Designer : Pierre-Yves Rochon
地点 Location : Paris, France
完工日期 Completion Date : 2007
摄影师 Photographer : Catherine Jaillon,
Claude Weber

LE PRE CATELAN

佩卡特兰餐厅

Le Pré Catelan first opened at the beginning of the 20th century and was a very popular place. A century on, the new design required an important aesthetic and intellectual thought. Pierre-Yves Rochon's aim was to give the place back its grandeur. Starting from the existing construction, he wanted to draw up a modern and striking concept that would emphasize the elegance of the original setting, in harmony with the surrounding nature.

This idea manifests itself through the furniture as well as the materials, the colours and the lighting. The predominant colours are green, black, white and silver. Evoking the greenery outside, green was used for the Salon Orsay's carpet and for the fitted carpet in the bar and rotunda. Black and white were used alternately to emphasize the differences between each area. The entrance to every room was designed in black in order to give an atmosphere of serenity, which brings out the impressiveness of the four main rooms designed in white and beige tones. This contrast emphasizes the space and architectural details.

Special attention was paid to the choice of furniture. Modern or classical furniture and mirrors in chrome and stainless steel contrast with the surrounding design, and reinforce the feeling of purity and light. The lighting too has been carefully conceived, with spotlights cleverly hidden to give the place an enchanting atmosphere.

佩卡特兰餐厅早在20世纪初期就已开幕，而且一直是很受欢迎、高朋满座的餐厅。此次全新的室内设计需要具有非常敏锐的美学与智慧思考。建筑师皮埃尔伊夫-罗逊的目标在于使这栋建筑重现其宏伟堂皇的气质。因此，设计方案的主导原则是从现存的架构着手，进而勾画出一个非常现代且令人惊叹的设计概念，以此来强调原有建构的宏伟。方案最大的挑战便在于如何构思出更具当代感并能与四周自然环境和谐共鸣的设计。

设计构思透过家具来呈现，当然也配合材质的运用、色彩和灯光设计。主色调是绿色、黑色、白色和银色。"奥赛厅"的地毯、酒吧区和"柔滩塔"圆形大厅的整片地毯都采用绿色，与餐厅外的一片葱翠绿意相连贯。黑色和白色的交替使用强调出厅内每个区域的相异性：每个厅房的入口皆设计成黑色以营造出平和稳重的气氛，同时也令人对设计成白色和米色调的四个主要厅房的印象更为深刻。这样的色调对比强化了宽阔的空间感及建筑的精致细部装饰。

餐厅内的家具经过精心挑选，无论是具现代感或古典风格，这些铬制、不锈钢制或镜面材质的家具，都与四周的空间装修设计产生绝美的对比，加强了简约纯净又明亮轻巧的设计信念。此外，照明设计也被特别构思，每个聚光灯都巧妙地隐身在造型含蓄的灯罩后面，更为整个餐厅营造出柔媚迷人的氛围。

设计师 Designer : Christian Ghion
地点 Location : Dubai, Dubai,
完工日期 Completion Date : 2008
摄影师 Photographer : Christian Ghion

REFLETS PAR PIERRE GAGNAIRE

皮埃尔-贾内的映照餐厅

Christian Ghion is more enthusiastic than he's ever been about designing Pierre Gagnaire's third gastronomic restaurant, in Dubai: "My goal was to design a place totally in tune with the chef's creative philosophy. To produce magic with a simple product, through human intelligence."

After crossing the spectacular lobby, you take the private lift down to sea level, where "Reflets by Pierre Gagnaire" is hidden, like a little gem. The restaurant is approached via a long corridor with a huge, mauve mirror lining one side, which reflects the shimmering, embroidered tapestry on the opposite wall. After passing a collection of glassware and candles shining in the low light, your attention is suddenly captured by an unusual sight on your right: through a cinematic frame in the mirrored wall you can look into the kitchen, where only the very best and sometimes the rarest ingredients in the world are assembled by magicians, performing for your own pleasure. The bustle in the kitchen contrasts with the serenity of the luxurious and intimate dining room, which you discover after you have passed the monumental wine cellar, standing tall like a giant aquarium housing the finest vintages.

A lush, purple carpet, lavender-coloured Murano chandeliers, custom-made rosewood Louis XIV-style chairs upholstered in pink velvet, padded gold banquettes in the alcoves, linen tablecloths and mother-of-pearl walls create an enchanting world. Christian Ghion has designed a harmonious space that successfully blends French tradition with modernity, which was exactly what Pierre Gagnaire wanted. You can choose to dine in the "private room" lined in lustrous bronze leather, where a secret little window hidden behind a mirror allows anyone who's curious to peek in at the pastry corner of the kitchen!

You can also choose to sit on the terrace below the waterfall. It overlooks the Marina and enjoys a splendid panorama of Dubai's most iconic buildings, most prominently the Burj Dubai, the tallest tower in the world, which glows in the sunset. A visit the powder room is an experience in itself! True to the theme of the restaurant (reflection), floor-to-ceiling mirrors play with your senses until you are totally disconcerted!

克里斯提安-吉翁对皮埃尔-贾内第三间高级美食餐厅的设计狂热简直是前所未有，他说："我希望勾画出一个能反映名厨贾内之创思的空间，用简单平凡的事物来创造非凡的魔力，但愿人类的智慧通达于此。"

沿着伸入迪拜海湾的狭长海岸，"皮埃尔-贾内的映照"餐厅隐身在新开幕的洲际酒店内，而酒店即坐落在名为"迪拜节庆城"的新兴城区内。一旦经过酒店华美壮观的大厅后，一部餐厅专用电梯将宾客载送到海平面楼层。一条神奇的长廊，一边是闪耀的绣花壁毯，另一边是一大面淡紫色的镜子，将一系列玻璃工艺品与蜡烛收藏品反映至无限深处。这长廊通至餐厅的主厅房，朝外看到的是玛丽娜街区以及建造中的迪拜歌剧院；而逆光向上望去，便是一片摩天大厦和目前名列世界最高的布尔吉-迪拜塔。在海岸另一头，呈现出一幕奇特又饶富诗意的景致，一家老旧的木船制造厂在淡淡的光线中隐约浮现，犹如过往的幽灵出现在令人惊叹的超现代城市中。

就在一片茄紫色的厚地毯前方，宾客会经过一个可看见厨房的大窗口，透过一个凿嵌在镜面墙上的窗框，呈现在客人眼前的是一个调理美馔的天地。一座壮观的酒窖犹如巨大的水族馆，收藏美不胜收的名葡萄酒，客人得绕过这酒窖才能进到餐厅入座。餐厅的主厅房里装饰了一盏盏富丽堂皇的穆哈诺玻璃吊、路易十四时代风格的座椅配上亮丽的粉红色绒布、隐密温馨的凹室空间、雅致的白麻材质桌巾，以及饰满来自深海的珍珠贝壳、令人赞叹的墙面，这种种巧思使得整个餐厅转化为一只珠宝盒般，绚丽而迷人。"私人厅房"全面铺上金褐色的皮质，还有一个仿佛隐藏在镜子后方的秘密小窗户，让最好奇的老饕能透过小窗观赏令人垂延三尺的甜点区。

客人也可以选择坐在露天座享用晚餐，露天座有一幕人工瀑布掩护着，可俯瞰大海美景。不过，无论如何，客人绝不能错过一个奇特的经验，去参观化妆室和通往化妆室那令人难忘的入口空间，从地面至天花板，一片片镜子装饰使空间格局产生错乱感，几乎令人迷失了方向！

设计师 Designer : Jean-Patrice Pham &
Afshin Assadian
地点 Location : Paris, France
完工日期 Completion Date : 2003
摄影师 Photographer : Djoon,
Philippe Santamaria

DJOON
迪容餐厅

Evoking the spirit of a loft, the Djoon combines a large, purified space, fitted out on two levels linked by a staircase, with impressive volumes highlighted by almost seven metres of ceiling height, natural light filtering through large plate glass windows, and daring architecture that plays with unrefined materials.

Here, several styles from different epochs cohabit, creating a pleasing sense of timelessness. The crude exterior called for the use of rich fabrics, colourful silks and velvets inside, in order to create a surprising contrast and to ensure a warm atmosphere. The contemporary, Italian style furniture is divided between cosy armchairs and divans, which warm up the concrete walls that have been decorated with a Baroque-inspired fresco.

The Djoon sees itself as a chameleon. While the cosy and welcoming space satisfies a business clientele at midday, at dinner time the Djoon dresses itself up with lighting. More filtered, the ambiance takes on a bolder identity and the decor looks resplendent under coloured lights. The designer's main priority was to give the Djoon a distinctive personality, a heart and a unique ambiance, and above all to find the subtle alchemy necessary to attract visitors without ever tiring them out.

迪容餐厅有着十分宽敞简洁、分散于两层楼的室内空间，上下楼层由一道楼梯结合，令人想到阁楼形式那种高大而开敞的楼中楼空间。它的面积宽广之外再加上约7米的天花板高度，自然光透过一大面落地窗照入厅内，见证着风格大胆的建筑空间与未经细致处理的简朴建材之间的趣味对话。

在此，多种不同时代风格的设计并存，呈现出一种令人愉悦、超越时空的感觉。设计师在建筑原有的简朴外观上搭配了多量的织布、高贵的料子和彩色绒布，以创造出令人惊叹的对比效果。同时也为空间带来温馨的气氛。这里设置的舒适扶手椅和软垫沙发是带有意大利设计风格的现代家具，为这个饰有巴洛克风格壁画的混凝土空间注入暖意。

迪容餐厅好比一只多采多姿的变色龙。中午时分，它以舒适温馨的空间来满足商业人士的用餐需求；到了晚上，餐厅换上五光十色的艳装，灯光更加柔暗、气氛更为热闹，厅内的装饰在彩灯的照耀下显得更灿烂。迪容餐厅的设计目标在于创造出一个与众不同的风格、一个独一无二的灵魂和空间气氛，尤其要为它注入那种能够吸引顾客一再光临的灵丹妙药。

设计师 Designer : Anégil
地点 Location : Paris, France
完工日期 Completion Date : 2005
摄影师 Photographer : Olivier de
Saint Blanquat, Anégil

ANGELINA

安洁丽娜餐厅

In 1903 the Austrian confectioner Antoine Rumpelmayer founded Angelina, named in honour of his granddaughter. Angelina soon became a celebrated meeting place, where Proust, Coco Chanel and others crossed paths in order to enjoy the famous hot chocolate and the incredible Mont-Blanc. In this context, the approach of the design agency (Anégil) had to be humble and respectful. The idea was to pay homage to this great lady who has lived through so many decades by giving her back the shimmer of youth.

The mezzanine and the two spaces at the back have rediscovered the lustre they once had. Treated with respect and in a spirit of continuity, these spaces are punctuated by reinterpreted allusions to the beginning of the century. Carpets with an "Angel Eye" pattern, a mirror that seems to float in the air and luminescent wall curtains reinforce the feeling of a precious world. A subtle play on the mouldings redraws the existing cupolas which had disappeared over the years.

The exercise consisted of integrating today's technology into an old setting: the air conditioning system, the lighting variation and effects, which have been approached with the greatest care, and details close to gilding for the treatment of the ventilation ducts, making them totally invisible. The room at the back has been treated like a reception room, its large dome adorned with a "Veronese trihedrons" chandelier, a marble chimneypiece, point de Hongrie parquet, and a warm and enveloping atmosphere.

Large carpets bordered with graduated darker colour intensify the feeling of thickness, of softness, of comfort, a perception strengthened by the parquet laths delimiting each zone. Fabrics with bronze and copper patterns resonate with the new colour scheme used for the blinds, the menus and the packaging. The leather of the medallion chairs brings light and reflections, conferring a special elegance on this legendary place.

奥地利糖果商安徒讷-洪佩勒梅耶先生于1903年创立了此午茶沙龙，命名为安洁丽娜，以纪念他的媳妇。自从一开幕，安洁丽娜即成为一个饕客绝不可错过的名店，大家慕名前来品尝著名的热巧克力和数不清的特色甜点，如白朗峰、普鲁斯特、香奈儿等。在如此辉煌的历史背景下，阿内基勒设计师事务所面对此方案的构思态度不得不谦恭谨慎。设计构想的主线在于礼赞历经多年考验的安洁丽娜，并为她添上青春活力的光彩。

在二楼及靠里面的两个空间都保留了旧时的吊灯。设计师以恭敬的态度、延续昔日装饰美感的手法来处理这些空间，因此在空间中时而出现20世纪初风格的设计，仿佛向那个时代致意。具"天使眼"风格的地毯、高高悬挂的镜饰，还有明亮的帘幕墙，强调出这个幽雅高贵的空间气氛。精致的线脚装饰巧妙地重新勾现出因历经多年而消损的圆顶。

此方案的重要任务在于如何将最新科技带入一个古老的环境空间。空调系统、照明设备的调整及灯光效果都以最谨慎的方式处理，跟金银细工吹炼珠宝那巧夺天工的技术一样。餐厅最里边的空间以作为宴会厅的功能来设计，宽阔的圆顶下饰了一盏维何内兹品牌独特的三角晶吊灯，大理石壁炉、斜角对称排列的匈牙利式木头地板，厅内气氛温馨迷人。

店内一大片地毯的边缘以渐层方式转为深色，更加强了地毯又厚又软又舒适的感觉，而且运用地板木条来划定每个区域的界限，也同时强调了地毯舒适的效果。青铜色和金铜色调的布料与店面的大帘子、菜单及产品包装设计这一系列全新的品牌标准色互相呼应。椭圆型靠背扶手椅的深色皮质微微反光，为厅内带来明亮的感觉，也为这传奇的午茶沙龙增添了独特的高雅气质。

设计师 Designer : Matali Crasset
地点 Location : Paris, France
完工日期 Completion Date : 2007
摄影师 Photographer : Toustem

TOUSTEM

图斯坦餐厅

Hélène Darroze, the chef from the South-West of France, fell in love at first sight with this setting, where the 13th-century stone walls and half-timbering recall the farmhouses of the Landes and inns of the Basque country. Three rooms, of which two are situated below ground, allow her to welcome 60 or so dining guests.

To bring this restaurant back to life, Hélène Darroze called on Matali Crasset, the artistic director of the whole project, and the architect Laurent Moreau. The mood is created even before you enter the restaurant, where the facade harbours bright and luminous colours: the orange and green that are emblematic of this chef's world.

Matali Crasset thought it a good idea to claim the space through the floor: an orange resin has been poured over it, seeming to splash in its wake over furniture, doors, plinths and even the plates! The restaurant's character is far from being constrained and the concept can express itself at will. In the cellar room, a false ceiling of stretched fabric covers the raw stone and breaks with the rustic and almost austere character of the place. The atmosphere is lightened, the room is illuminated and transformed. The very contemporary white furniture answers the sombre beams.

Finally, Hélène Darroze had fun by sliding very personal touches into the decoration: an impressive bright pink light covering at the entrance, a dark wood counter and high stools that are perfect for enjoying tapas and cocktails, and a massive spit on which pork and poultry are roasted, one after the other...

此餐厅的诞生是因为爱莲娜-塔后姿，这位来自法国西南方的主厨对这个地方一见钟情，这里有建造于13世纪的石墙，而且木材梁柱结构外露的柱筋式建筑让她联想到法国西南部龙戴地区的农庄和巴斯克地方的旅店。此地拥有三间用餐厅，其中两间位于地下室，可以容纳60多位宾客。

为了让这个地方重生，爱莲娜-塔后姿找来了玛塔丽-葛拉瑟负责整个设计方案的艺术指导，加上建筑师罗鸿-莫候的合作，目的要让这历史悠久的老房子再度充满生气。客人一到门口就可感受到餐厅的设计主调，外墙运用鲜艳明亮的色彩，这些色彩正是主厨所熟悉的世界中最具象征性的橘红和鲜绿色。

设计师玛塔丽-葛拉瑟认为，本方案必须从地面的处理来掌握整体的空间，因此她采用了橘红色的树脂来作为铺地的材料，鲜艳的色彩于是一路泼洒到桌椅、家具、厅门、踢脚板，甚至蔓延到餐盘上来！这个餐厅的装潢布置并非就此一成不变，它的设计基础概念能够依主人的兴致所至，而延伸出其他多种变化。在地下室里，以绷紧的帆布延展形成的天花板覆盖于石墙之上，打破原本空间极其朴素甚至过于严谨的风格，进而使得整体的空间气氛变得更轻松，厅内变得明亮而焕然一新。此外，极具现代感的白色家具和暗色系的结构梁也产生对比的空间趣味。

最后，爱莲娜-塔后姿更是尽兴地将其深具个人色彩的风格注入这里的装修设计，像是入口处令人眼睛一亮的粉红色鲜艳灯光照明、深色原木的吧台和高脚椅，让顾客在此品尝鸡尾酒和西班牙小点心、享受从大型串烤机那儿传来的阵阵烤鸡和乳猪香味……

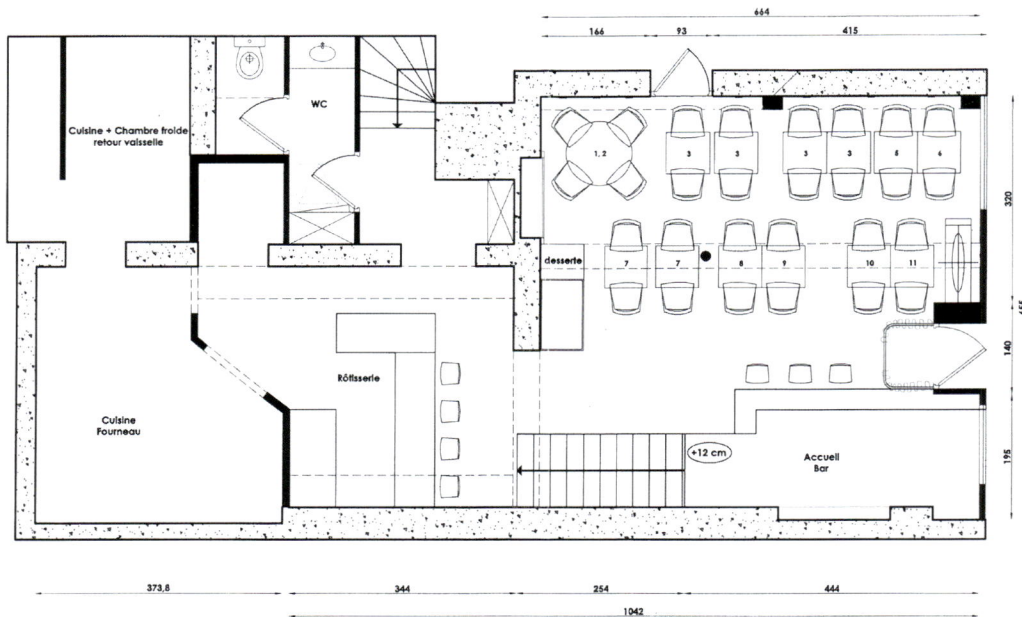

设计师 Designer : Frédérique Gormand &
Christophe Vendel
地点 Location : Paris, France
完工日期 Completion Date : 2008
摄影师 Photographer : Jean-Charles Valienne

L'EPI DUPIN

艾皮杜庞餐厅

The Epi Dupin is a small local restaurant (50m²) established 14 years ago, where habitués and an international clientele come together over the inventive and subtle cuisine of chef François Pasteau. Modest though it is, the restaurant is a popular gastronomic haunt.

To preserve the "Parisian" spirit of the place, its typical architectural features such as the half-timbered stone walls and beams have been left visible. This gives the space a certain modernity while keeping it in line with a sense of tradition and the culinary inventiveness on offer.

The main project was to move the bar to the back of the restaurant in order to group together the practical areas and to free up the reception space on the facade side. This allowed the owners to offer a different, less formal kind of welcome by creating a communal table as a prolongation of the bar. The continuity of "communal table-bar", with the masses in good proportion to each other, also allowed them to give the place its own character. Finally, a bottle rack turned into a banquette completes the ensemble.

The materials were chosen with simplicity and rigour in mind, setting up a contrast between the warmth of solid wood and leather, fine and natural materials, and the black synthetic resin that forms the bar, the communal table and the bases for the wooden table. On the ceiling, an acoustic fabric softens the noise levels. The beams have been painted white to make them less dominant and to give luminosity to the place.

艾皮杜庞是一间街区小餐馆，面积仅50平方米。餐馆开立14年来，多是熟悉店家的常客和从世界各地慕名而来的老饕，齐聚一堂来品尝弗朗索瓦-帕思图主厨极具创意又精致的美食。这是一个外表简朴的餐馆，但客人皆闻其美食之名而来。

为了保留餐馆内原有的"老巴黎"味道，其典型的建筑结构，如石墙和木材梁栏结构都保持外露。如此一来，传统的空间风格搭配创新的美食，反而给予餐馆某种鲜明的现代感。

设计方案主要的重点是先将吧台移至餐馆最里面的位置，把工作空间合并在一起，也因此空出靠店面外边的位置以作为接待客人的空间。这样的格局变动能够提供一种截然不同的接客方式，比较轻松，避免刻板，并留出空间摆置一张大主宾桌，延伸了吧台空间。"主宾桌-吧台"这条连续动线构成一个看起来高朋满座的整体空间，也因而成为此餐馆的特色。加上一组酒架下方转变成长条形软垫椅，使整体空间的利用更为完善。

厅内所运用的材质呈现出简洁又严谨的风格，有暖色调的实心木与皮质的对比，有高贵自然的材质，还有黑色的合成树脂，用来制作吧台和主宾桌，也用来镶接木制的餐桌。天花板上安装了一块隔音布以消减噪音。外露的木梁被漆成白色以免过度抢眼，同时也让室内整体看起来更明亮。

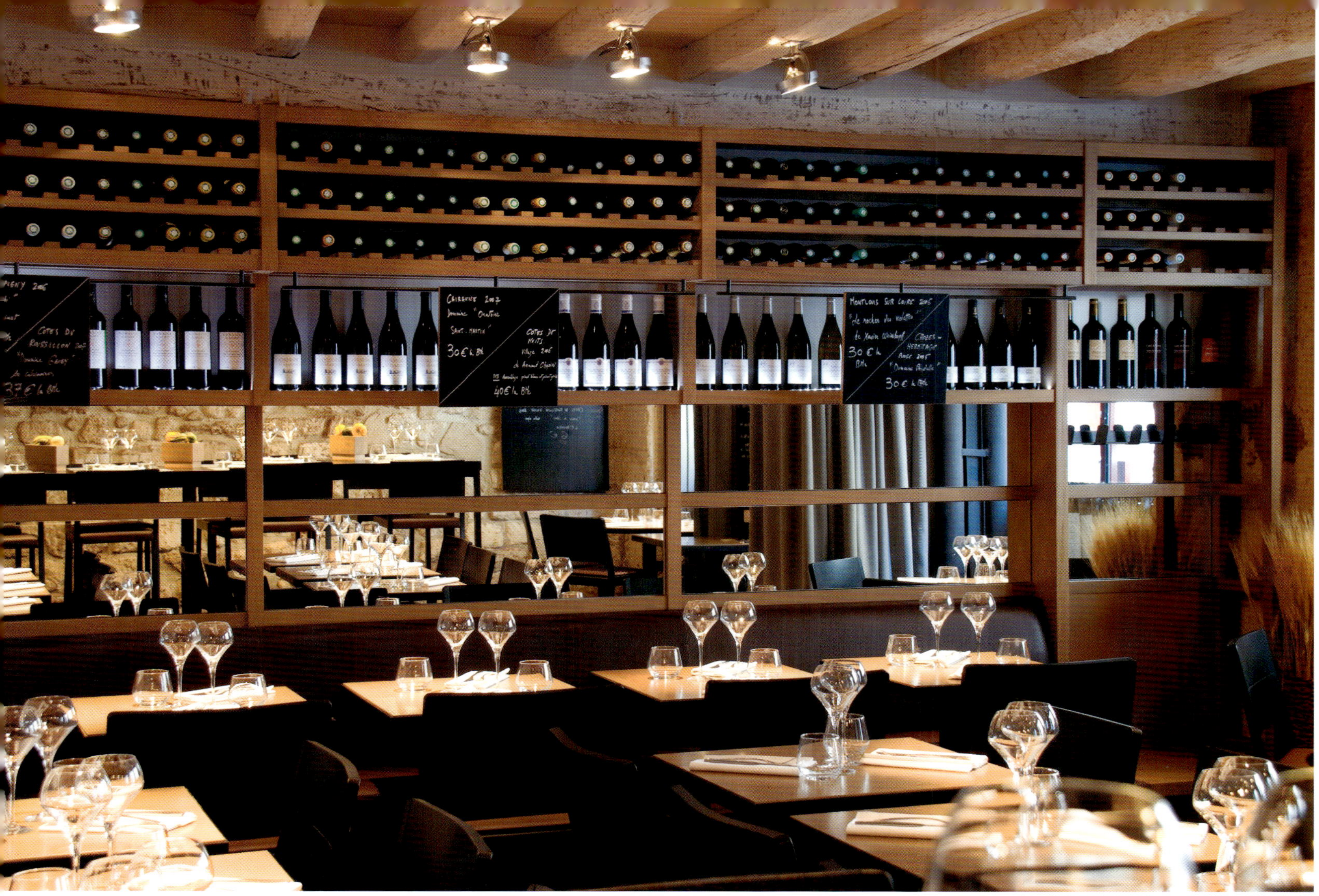

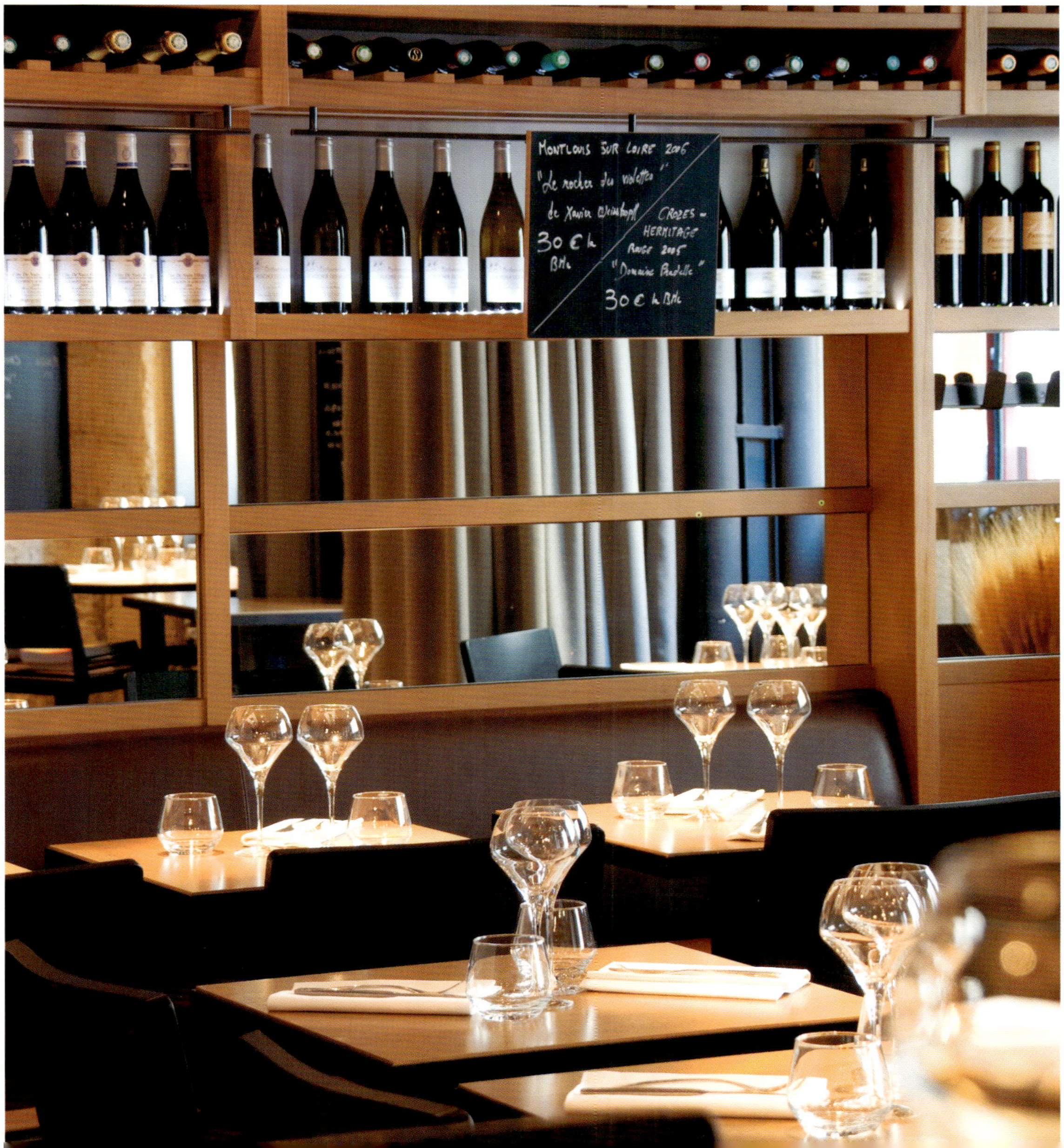
MONTLOUIS SUR LOIRE 2005
CROZES-HERMITAGE
ROUGE 2005
30 €

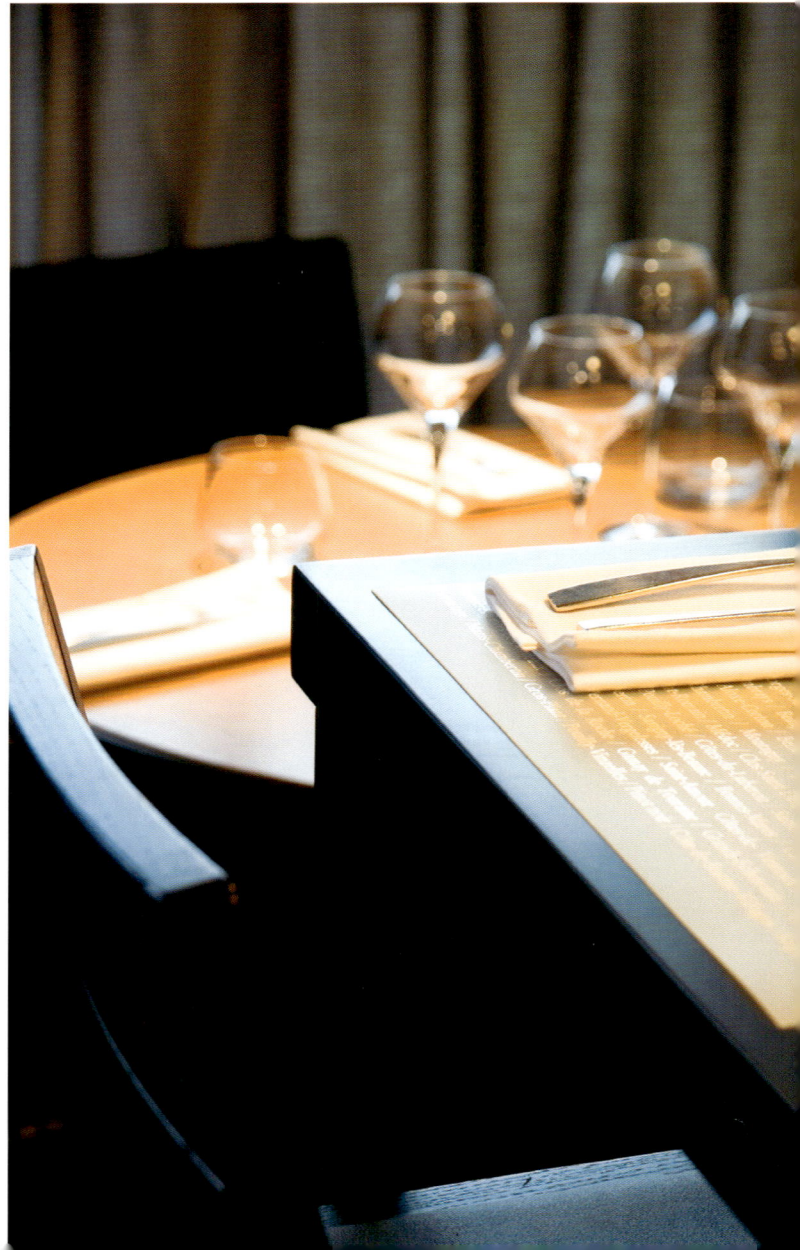

设计师 Designer : Ralston & Bau
地点 Location : Paris, France
完工日期 Completion Date : 2003
摄影师 Photographer : Wijane Noree

SUR UN ARBRE PERCHE

栖息树上餐厅

Sur un Arbre Perché is a story that is told while eating... Its inspiration naturally comes from the fable of Jean de la Fontaine. Paris is an astonishing and marvellous city, and as such is teeming with movement and stimuli. The restaurant's designers wanted to create an environment that would make its visitors feel really cared for... An authentic and natural space that would provide a peaceful pause in the midst of the commotion of the city.

The idea of the fable came up as something natural, simple and obvious. To give a form to childhood dreams, without compromising on the materials and the way they are perceived: this is why the perched cabins are made from authentic old Swedish barns that are more than 200 years old. Dismantled with care, they were used to construct all the "Nests" of Sur un Arbre Perché. To give the diners privacy, the Nests have been placed at different heights; each diner can choose his or her branch, softened with fat and comfortable cushions.

More playful visitors can also enjoy their delicious meal on a swing! And those who come with small children can have a Nest fitted with a banquette adapted to a child's size. The air of the restaurant is filtered to leave urban impurities outside. Calm, replete, each guest can finish his feast with a shiatsu massage. Beyond merely enjoying food, this restaurant offers a complete relaxing experience.

"栖息树上"是个要坐在餐桌旁边吃边讲的故事……其设计灵感显然来自法国作家拉封丹的寓言"乌鸦与狐狸"。因此，面对巴黎这个令人惊叹、不可思议、充满活力和刺激的城市，餐厅的设计师们希望创造出一个能够真正关照到客人的环境，为他们构想一个独特、自然的空间，在热闹喧嚣的城市中营造出一个可让人放松享受的天地。

叙述一个寓言故事的设计灵感便如此自然而然、理所当然地产生。要建构出孩童的梦想，绝对要选用最适当的材质并重视其感官效果；因此，厅内所有架高的棚子都是由真正的旧谷仓建材而成。这些来自瑞典北方的古老木头已有两百多年历史，得先小心翼翼地将它们拆卸下来，再用来制造"栖息树上"餐厅内所有的"窝"。为了让客人保有各自的亲密空间，每个窝棚都架在不同高度上，每个人可以选择自己要栖息的树枝，窝里放满了舒适的大靠枕。

若客人童心未泯，可以选择坐在秋千上享受他的美食！如果有孩童伴随，有个设计了适合小孩身形的软垫椅。厅内的空气还经过调节过滤，滤掉外面城市的乌烟瘴气。客人还可在吃饱喝足后，心情放松时，来个日式按摩；除了愉悦地饱餐一顿，当然也要让精神更加焕发才行。

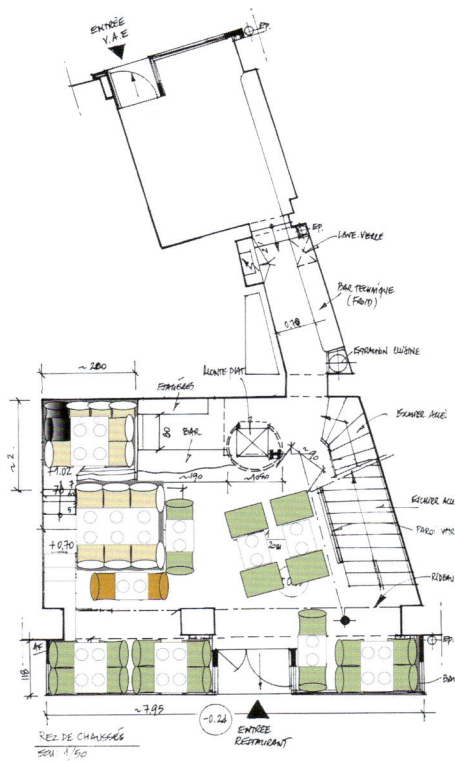

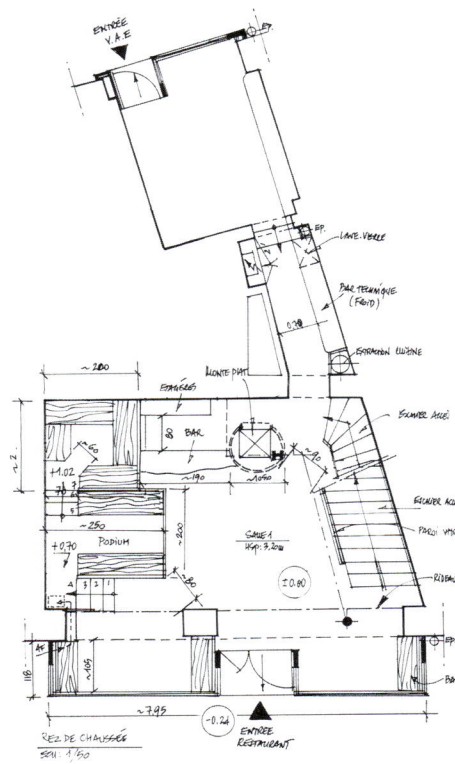

设计师 Designer : Armand &
Martine Hadida, Piero Fornasetti
地点 Location : Paris, France
完工日期 Completion Date : 2008
摄影师 Photographer : Patricia Belair

L'ECLAIREUR

先锋餐厅

The decor of the bar and restaurant of the Paris boutique L'Eclaireur is a transposition of the world of the Italian designer Piero Fornasetti. Decorated by Martine and Armand Hadida in collaboration with Barnaba Fornasetti, it takes its inspiration from the Dulciora cake shop in Milan, designed by Piero Fornasetti in the 1950s. Two terraces, one in the Galerie Royale and one on the rue Boissy d'Anglas, complete it.

At the entrance, the gold leaf decoration known as "Grand Coromandel" is inspired by Chinoiserie even though its motifs are neo-Classical. Near to the bar is found a series of printed wood panels under which one can sit on stools with red lips, or on reinvented seats. These themes echo the original erotic drawings of Fornasetti, visible in a "naughty" corner. A folding screen by Fornasetti also creates a feeling of intimacy here.

Drunken monkeys and choux pastry towers for the bar, card game motifs inspired by the surrealist metaphysical landscapes of 20th-century Italian art, "madrepore" Medusas for the restaurant, or sun and moon symbols (very important to Fornasetti), and Adam and Eve, etc., form a fantasy world. In the restaurant, trompe l'œil marble columns simulate a window onto a landscape of cards and butterflies flying above a bookcase. A ladder repeats the motif of the "Scaletta" screen, two plates from the series "Theme and Variations" appear, as well as vases inspired by the decoration of the Dulciora cake shop, and the whole is overhung by a frieze of owls on a midnight blue background. A decor designed for dreams in a very individual space.

玛婷和阿蒙-哈迪达两人与芭娜芭-福纳瑟提合作完成先锋餐厅的室内装修，设计灵感来自20世纪50年代皮耶赫-福纳瑟提在米兰设计的一家名叫杜修哈的糕饼店，餐厅内吧台空间和用餐区的装修设计完全是这位意大利设计师的创作世界的转移。餐厅还有两个露天座餐饮区， 一个位于皇家廊内，另一个靠在玻西-东阁拉路边。

餐厅的入口空间有一件名为"宏伟的科若蒙德尔"的作品，是一件深具中国风味的金纸装饰品，但是其上的图案属于欧洲新古典风格。靠吧台的墙上有一系列木制绢印画作，客人可坐在红唇形的矮凳上或创意新颖的椅子上用餐赏画。这些装饰的主题都受到福纳瑟提的情色素描原作的影响，尤其在称为"调情"的餐厅一角更是明显，一座福纳瑟提所创的屏风也为这个角落营造出私密的空间气氛。

吧台区的装饰主题是醉猴和糕饼塔，用餐区多是一些受20世纪意大利艺术中超现实抽象风景画影响的扑克牌图案，或是石珊水母，或福纳瑟提最喜欢的太阳和月亮，或亚当和夏娃等形象，整体犹如一个神话幻想世界。在餐厅区的墙面上，运用错视虚拟的画法呈现出一根根大理石柱，逼真地画出风景明信片般的窗外景色，蝴蝶在书柜上翩翩起舞。有一个梯子是拟摩"斯卡雷塔"屏风中的梯子画法，两个源自"主旋律与变奏曲"系列作品的餐盘，还有一些花瓶，其创作灵感来自杜修哈糕饼店的装饰，而且这些壁画最上面还有一道长形壁画呈现出蓝蓝夜空中的猫头鹰。这里的一切装饰引领人进入梦想天地，真是一个独特的餐饮空间。

设计师 Designer : Donatelle Piana &
Philippe Batifoulier
地点 Location : Lyon, France
完工日期 Completion Date : 2008
摄影师 Photographer : Alain Rico

LA TASSEE

天时餐厅

The architectural concept of La Tassée is based on the idea of restoring pride in the historical referents of this old Lyonnaise institution that has been in the same family for three generations.

The timeless colour scheme of browns, taupe and sand was thus retained, and given new harmony through sober and elegant lines. The walls alternate between wooden panelling inset with architectural lighting, and a pony skin effect. The resin floor gives a contemporary look to the setting and is echoed in the drop ceiling, into which are set chandeliers that have been customised with silver wine-tasting cups, a symbol of the restaurant.

The restaurant's historic Bacchanalian frescoes have been brought back to life like Old Master paintings through the addition of a frame and elegant lighting. The innumerable personalities who have frequented La Tassée over more than 60 years now form part of the decor: their photographs, tinted with the colours of the frescoes, are inserted into a Perspex wall.

The bar space has been laid out with alcove banquettes specially designed for La Tassée, which give it a cosy atmosphere. Italian furniture brings a final touch of refinement and comfort.

天时餐厅的建筑设计构想专注在发扬这间目前已由第三代子孙在掌厨的古老里昂餐馆的历史背景。

因此，设计师采用了一套不退流行的色彩：棕色、深黑色和沙土色，配合高雅简洁的线条，呈现和谐的整体。墙上交替呈现镶嵌建筑式灯光的木制墙饰和小马毛皮的材质效果。树脂地面赋予整体空间一种极具现代感的气质，并与天花板横梁的突出结构相呼应。在这天花板上嵌着一盏盏利用品酒小银杯集制而成的吊灯，这独特的吊灯成为此餐厅的象征。

描绘酒神与历史的壁画经过裱框与优雅的投射灯装置，呈现出大师杰作般的价值感。60多年来常来餐馆的众多各界名人从此也成为餐厅装饰的一部分：他们的照片套上符合壁画的各种色调，被压裱在有机玻璃制的墙板上。

酒吧区放置了一排排特别为天时餐厅设计的圆弧形软垫沙发，营造出十分温馨的气氛。这里的意大利家具也为餐厅添加了精致舒适的设计笔触。

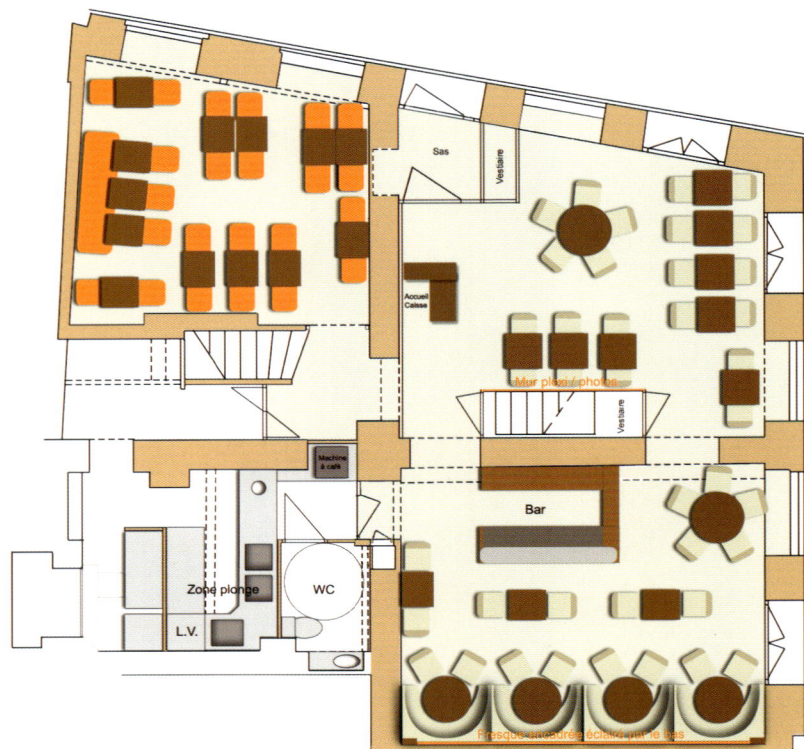

设计师 Designer : Saguez & Partners
地点 Location : Paris, France
完工日期 Completion Date : 2007
摄影师 Photographer : Olivier Seignette,
Mickaël Lafontan

CARRE MONCEAU

梦梭四方餐厅

An intimate ambiance is established from the minute you enter the restaurant Carré Monceau – an atmosphere both disciplined and welcoming, like a select English private club. A choice of different spaces is offered to the visitor. In the lounge you can wait for friends to arrive while drinking an apéritif. At the bar, you can eat informally, standing up. Square and round tables in the dining room can seat ten, and groups of six to twelve people can sit in two private rooms.

Opening up onto the dining room, the butlery is a work space and window onto the kitchen. Three open corridors give a glimpse of the "atelier", the den of the art of cooking. On the other side, the room gives onto the central patio. Large plate glass windows let natural light flood the restaurant while, in the same spirit of transparency, a lower band of frosted glass brings just the right amount of discretion: to see without being seen.

The spirit of the place is inspired by the Vienna Secession, a culturally rich artistic movement whose personalities included those of the "Plaine Monceau", artists of the 19th-century European revival who introduced the Modern Style. Two head-to-toe portraits of the Comte de Montesquiou, a businessman and dandy, have been placed at the end of the room, at the entrance to the private rooms. With his discipline and elegance he embodies the European spirit of the epoch: black and white, enlivened by a touch of "Vienna Secession" yellow…

一进梦梭四方餐厅就可感觉到一种亲密的氛围，既严谨稳重，又热情温馨，如同那些英式的上流社会俱乐部。厅内提供客人几个不同的空间：沙发酒吧区邀客人喝一杯等待入座；在酒吧区，人们可以轻松地站在小吃台旁用餐；餐厅主厅房的方形桌和圆形桌可容纳10位宾客；而6～12人的小团体可分别独享两间私人厅房。

工作人员的配膳空间与厨房的窗口部分很大方地呈现在餐厅客人的眼前。客人可透过三个开启的通道瞥见"工作室"，那创造美食艺术的神秘圣地。餐厅另一侧朝向中庭，日光透过宽大的落地窗映照进来，溢满整个厅内。还有那一片片长条形的亮雾玻璃，也呈现了透明美感，更为此空间带来恰到好处的隐密感，可观望出去却不被外人看到。

此餐厅的内在精神受到维也纳分离主义时期的影响，那时期文化蓬勃发展，丰富多元，很多名人都参与当时活动，包括梦梭平原派的人士，共同致力于19世纪欧洲艺术的革新，为它带入现代风格。两幅孟德斯鸠伯爵（花花公子名商）的全身画像挂在餐厅最里面，矗立在私人厅房的门边。他代表了当时欧洲的人文精神，体现一种严谨优雅的气质。因此餐厅以黑色与白色为主调，再饰点"维也纳分离主义"那独特的黄色。

设计师 Designer : Philippe Boisselier
地点 Location : Paris, France
完工日期 Completion Date : 2007
摄影师 Photographer : Silent Factory,
Philippe Ruault

LE SAUT DU LOUP

防狼堑壕餐厅

The restaurant Le Saut du Loup is accessible by two entrances, one via the Musée des Arts Décoratifs and the other through the Carrousel garden. In garden design, a "saut de loup" – or in English a ha-ha – is an invisible limit, made up of a ditch and a wall leaving the perspective intact. Used as protection, it allows for perfect continuity between two landscapes. The whole of Philippe Boisselier's project is linked to this idea of a "saut de loup".

On two open levels between the museum and the garden, the project plays with the contrast between black and white to construct the space. The first floor with its black ceiling and white floors answers the ground floor with its black floors and white ceiling.

The works with materials such as Corian, the play of light and graphic touches, contribute to the modernity and discreet luxury of the place. Silvered mirrors in the aperture of the windows capture the reflections of the city, multiply the volumes and maintain the constant dialogue between the interior and the exterior.

The lines of the single-armrest banquettes in black wool (Kvadrat) designed by Philippe Boisselier sort well with the grey, black or white tables; the light, round tables designed by Charles Eames; the square armchairs in grey leather and the Miura bar stools by Konstantin Grcic.

"防狼堑壕"位于巴黎市中心，可从瑞弗利路装饰艺术展览馆入口或卡鲁榭花园方向进入餐厅。在园林艺术领域中，"防狼堑壕"指的是一道看不见的界限，由壕沟与堑墙组成，留下完整无碍的透视远景。这保护性壕沟使两个景之间能保有完美的连贯性。菲力普-波瓦瑟里耶的整个方案即是依循这著名的"防狼堑壕"的建构原则。

这个两层楼餐厅位于装饰艺术馆和花园之间，方案设计以黑白对比来建造整体空间。一楼采用亮面黑石地板及白色天花板，与二楼的雾面白石地板及黑色天花板对应。

不论是材质的运用（如"可丽耐"人造面板）、灯光的效果或图案设计都使餐厅呈现出既摩登又含蓄高贵的气质。墙上窗洞设置了银光闪闪的镜子，捕捉着城市的映像，增加了空间深度感，也让餐厅内部与外界不断地进行对话。

由菲力普-波瓦瑟里耶设计的黑色羊毛呢绒制（科法达品牌）的单扶手长软垫椅被摆置成一排排直线，巧妙搭配查理-依姆斯设计的灰色、黑色或白色餐桌以及色彩浅亮的圆椅，也与康斯坦当-戈瑞克所设计的灰色皮质方形扶手椅及米欧哈系列高脚凳产生协调一致的美感。

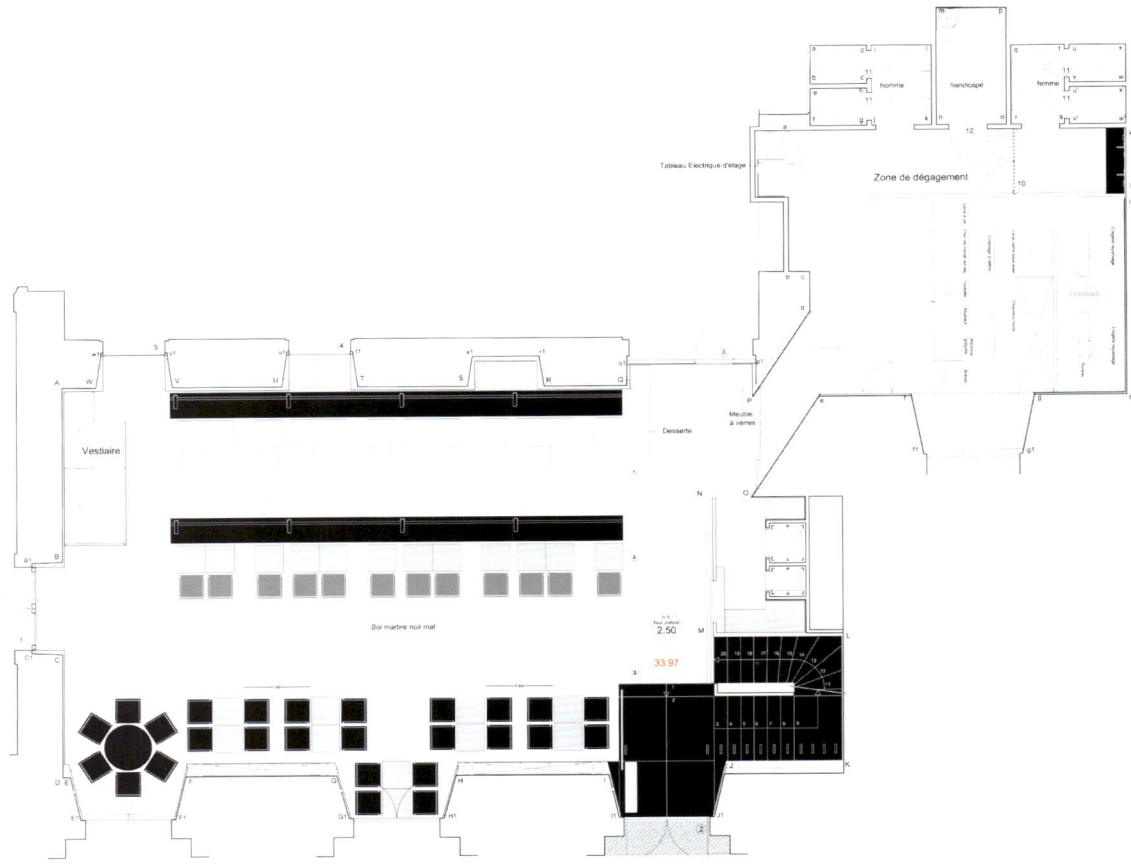

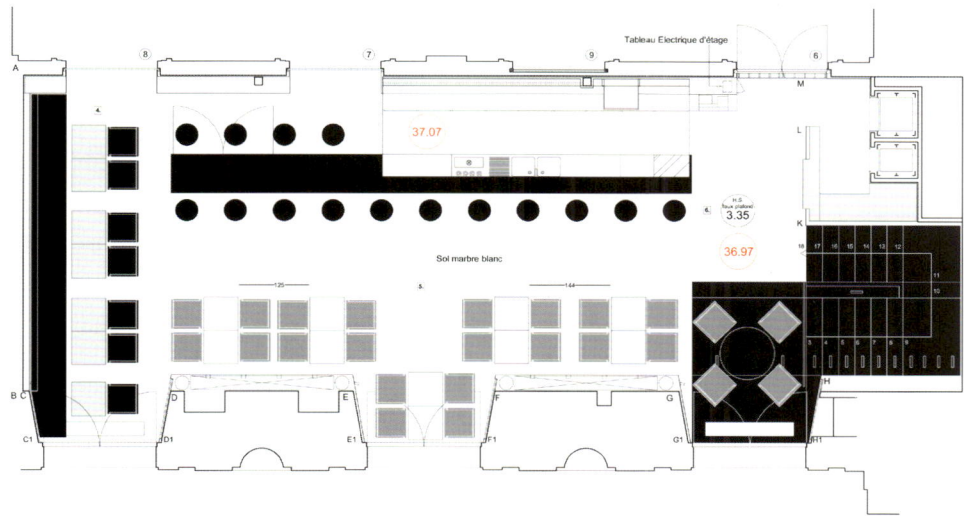

设计师 Designer : Sybille de Margerie
地点 Location : Courchevel, France
完工日期 Completion Date : 2006
摄影师 Photographer : Marc Bérenguer

LE RESTAURANT 1947

1947餐厅

In Courchevel, in the hotel Le Cheval Blanc created by Sybille de Margerie, the restaurant Le 1947 makes a daring statement, in harmony with the surrounding nature. A refined world, it rejoices in a play of shadows and light with the mountains as its backdrop... Details ranging from stone slabs in an Opus Romanus pattern, a fireplace, bronze doors patinated gun-metal grey, the quality of the skirting and ceilings in gingerbread-coloured walnut to the door lintels in alabaster all show to what extent talent has gone in to the choice of materials and colours.

The curtains are in striped grey, black and bronze taffeta, the seats with their smooth lines are covered in black and lichen bast velvet with the backs in striped taupe, grey, bronze and oatmeal velvet. Bronze and alabaster wall lights and lamps with silk shades diffuse a subtle light. The contemporary chimneypiece, whose hearth is at the height of the tables, also adds to the spirit of the place.

The same colour palette is carried through to the table settings. Taupe and white plates have been designed by a well known house for the Cheval Blanc. Smoked glass tumblers, silverware and knives with horn handles and tablecloths with a taupe embroidered border over an under-cloth in snowy-white velvet offer delicacy and strength, the true spirit of this place and of a winter in Courchevel.

1947餐厅位于高雪维尔城，在席比勒-德-马格丽设计的白马酒店内，餐厅的整个构思十分大胆，却又与四周自然环境协调呼应。这个设计极其精致的餐厅当中富含着无数的巧妙光影效果，而且拥有珍贵的山岳景致。室内设计呈现出丰富的细部巧思，例如：具有特殊纹路坋压花石材、壁炉、黑灰色光泽的铜门、高质量的建筑底座、蜜糖蛋糕色的胡桃木天花板、以雪花石膏精制出来的门侧，这些细节一再显示出设计师在材质与色彩运用上的高度掌握能力。

塔夫绸制的窗帘饰上灰、黑及金铜色条纹，线条简洁的座椅铺上黑色和苔藓色坋绒布，座椅背面则饰了深黑色、灰色、金铜色和麦黄色的直条纹。以铜与雪花石制成的壁灯及饰有丝质灯罩的落地灯散发出柔和柔美的灯光。壁炉的设计极具现代感，其炉膛跟餐桌的高度相当，符合厅内整体的感觉，使之更为协调统一。

这些室内设计风格也延用至餐具的设计。白色配深褐色边的餐盘是由一个著名厂家特为白马酒店设计的。烟玻璃制的水杯、银制餐具和角形刀柄的刀子、饰有深褐色边排线条的桌巾，桌巾下面又铺上一块雪白的绒质桌布，整体餐桌的摆设呈现出既精巧细腻又强而有力的精神，这也就是此地的精神，盛邀宾客在高雪维尔城渡过一个难忘的冬季。

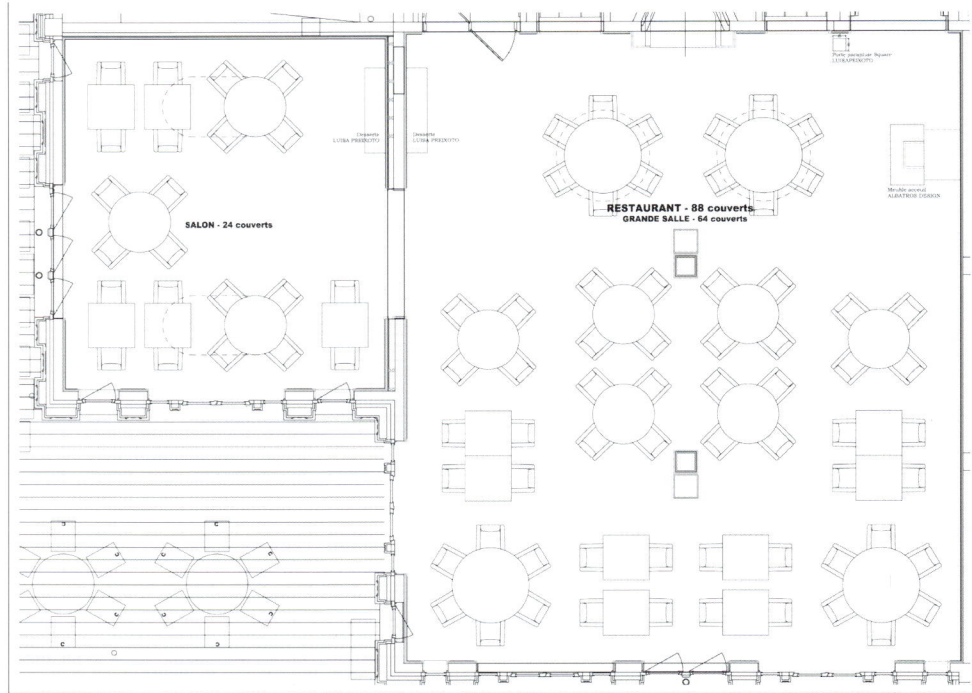

SALON - 24 couverts

RESTAURANT - 88 couverts
GRANDE SALLE - 64 couverts

设计师 Designer : Olivier Gossart
地点 Location : Paris, France
完工日期 Completion Date : 2006
摄影师 Photographer : Jacques Gavard

LE BOSQUET

波斯科餐厅

The brasserie Le Bosquet is situated on avenue Bosquet, at the foot of the Eiffel Tower, between the Invalides and the Champ de Mars. The fruit of a meeting between the owner of the restaurant, Jean-François Trocellier, and the designer Olivier Gossart, the project was designed to respond to a new international clientele that is receptive to the French lifestyle and to innovation.

The name "Le Bosquet", so evocative of nature, served as the main theme from which to design an elegant and homogenous whole using fine materials such as leather and wood with the same background colour of "earth", between beige and taupe. This warm environment provides a setting for plants and trees such as olives, maples, Christmas trees in the window and a bar made from a collection of small birch trunks.

An adjustable lighting scheme allows one to change the ambiance. Long ceiling lights in carbon fibre in the shape of bottles are visible from the outside and define the brasserie visually. Projected on the wall are artworks or artists' films. The furniture designed by Olivier Gossart uses different leathers, metallic fabrics, wood and stainless steel. Elsewhere the designer called on the artist Marc Hayraut to create a photograph of vegetation, imaginatively worked in the form of strata to match the concept. This image has been elegantly repeated on Le Bosquet's menus.

波斯科餐厅位于巴黎波斯科大道上，介于荣军院和战神广场之间，紧邻埃菲尔铁塔。餐厅老板让佛朗西斯-投瑟里耶和设计师奥利维-葛萨两人投缘的相遇，促使了此餐厅构想的成形与实现。此方案的目标在于满足国际性新客群的需求，吸引那些想深入了解法式生活艺术并且探求创新前卫品位的饕客。

餐厅取名波斯科（Bosquet，法文意谓：树林），令人联想到大自然的意象，于是呈现"树林"便成为餐厅构思主线，以设计出幽雅和谐的整体空间。设计师利用一些高贵的材质，如皮料或木材等，处理成一个具有"土地"色感的空间背景，介于淡土褐色与暗褐色之间。这个具有温馨色调的环境衬托出厅里植物树木的美感，就像那些被奉在橱窗里的迷你橄榄树和小枫树，还有那运用一块块桦树树干堆叠成的吧台。

可调整变换的灯光能为餐厅塑造出不同的气氛，而从餐厅窗外就看得见的碳纤维瓶形长吊灯，则成为此餐厅令人印象深刻的特色。厅内的墙上投影放映着艺术家的作品或影片。由奥利维-葛萨设计的家具采用了各种不同的皮料、呈现金属光泽的布料、木材和不锈钢。除此之外，奥利维-葛萨还请艺术家马克-海何创作一幅植物摄影作品，表现出"地层"的形式概念，而这地层的概念也巧妙地重现在波斯科餐厅的菜单设计上。

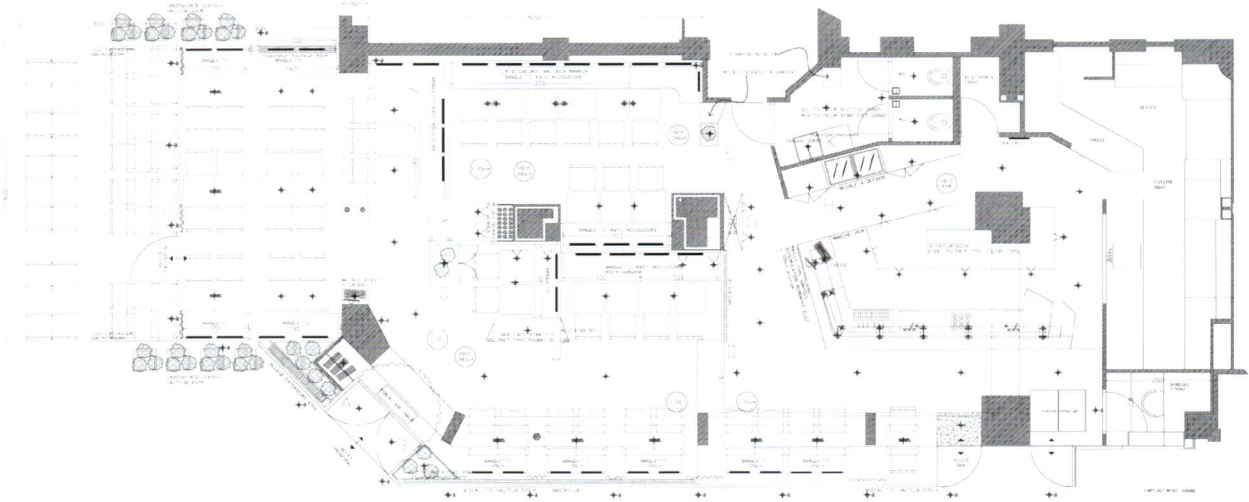

LE BOSQUET

设计师 Designer : Jean-Philippe Nuel
地点 Location : Crozet, France
完工日期 Completion Date : 2008
摄影师 Photographer : Jiva Hill,
Pierre-Dominique Brunet,
Anne Vachon

SHAMWARI

沙慕里餐厅

The restaurant and its bar are distinct from Jiva Hill Park Hotel, with their own identity which looks to combine a feeling of the mountains with a refined sophistication. Here you will find a contemporary spirit with a particular studied elegance, creating an intimate atmosphere brought about, among other things, by the use of black.

A fireplace, and numerous suspended lights made up of candles that reflect in the mirrors, give the whole place a magical aspect. Warm browns are enlivened from time to time by the armchair "Luca Bridge" in a bright yellow fabric and by the reflections generated by the varnished and chrome furniture. These go together to create a contemporary setting, placed in the middle of nature with large plate glass windows and a wide terrace. The relationship with the outside is an important element, alternating complete transparency with a play on screens.

A made-to-measure wine cellar allowed the designers to create a feature while organizing the space. The bar has been conceived as an enlivening visual element which offers a less conventional way to eat: it is the height of a normal table – lower than a classic bar so as not to form an obstacle in the space.

沙慕里餐厅酒吧与所在的季瓦山酒店风格不尽相同，它希望将山岳的精神融入一种简洁纯雅的设计之中，展现出自己的特殊格调。餐厅处处呈现着极为现代感的设计，并且运用巧思来营造出温馨亲密的气氛，例如对黑色调的使用。

一座壁炉，加上一盏盏蜡烛吊灯，交相辉映在镜中，赋予整个餐厅一种仙境般的感觉。温暖的棕褐色调偶尔点缀上铺饰鲜黄布料的路卡桥系列座椅，也映着亮漆和镀铬家具反射过来的光影。厅内整体与大落地窗及宽广的露台，共同创造出一个被置于大自然中的时尚空间。餐厅内部与外界之间的互动关系是设计的要素之一，两者之间可以完全透明相通，也可以运用栏栅的不同关启程度而带来变化。

依场地量身定做的酒窖为空间带来动感，同时主导着空间格局。吧台区也被当成一个活动空间元素来设计，并且可提供客人一个异于常规的午餐空间：它的高度被设计成与正常餐桌相同，因此也比一般的吧台低矮，避免成为空间中的阻碍。

CAFE RESTAURANTS

酒吧-简餐-下午茶馆

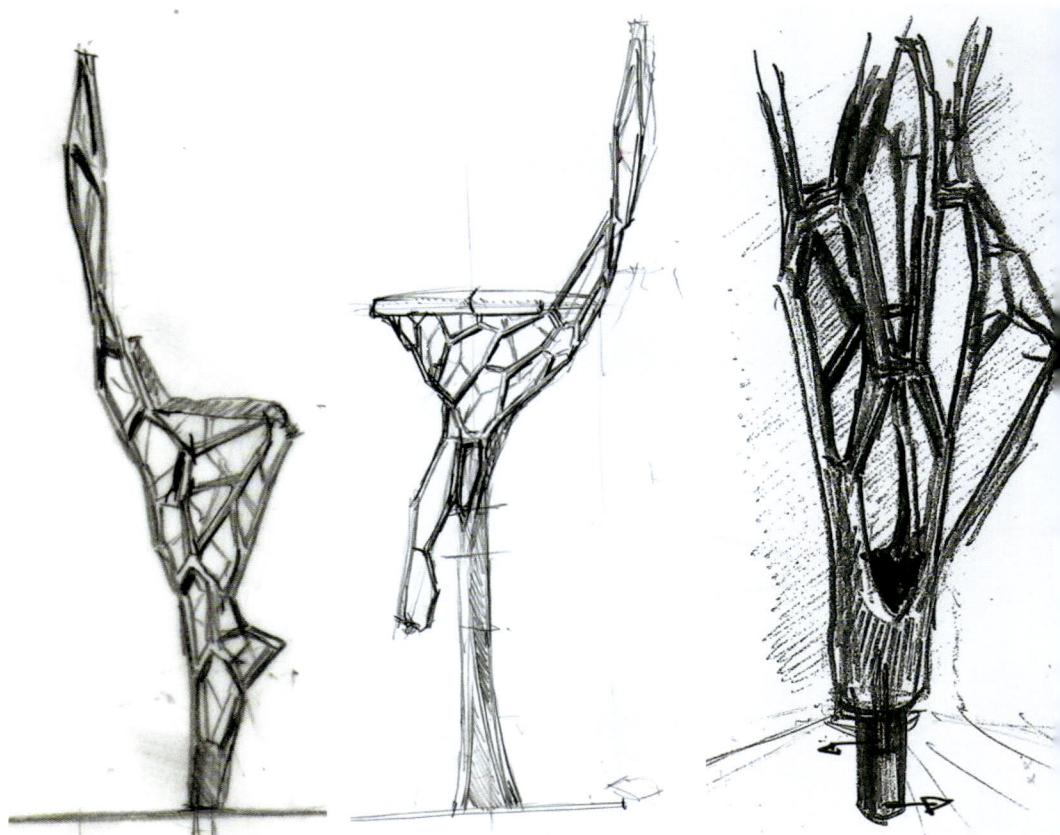

设计师 Designer : Roxane Rodriguez
地点 Location : Paris, France
完工日期 Completion Date : 2008
摄影师 Photographer : Thierry Malty

LE LADUREE BAR

拉杜蕊酒吧

What strikes one on entering this den or hideaway, set back from the Champs-Elysées, is the dense mesh forming the framework of this enigmatic place where time stands still.

The hieratic chairs seem to form the coral cladding of a majestic sunken ship... At the same time, it would also be true to say that this vast, organic, alveolar structure has come together like a chrysalis hiding the gestation of an imperceptible/invisible world, between two infinities... Let's look a little closer at these chairs. The seat, a purple cluster, forms the primordial chaos, the foundation, of this expanded aluminium structure. The following structure, where each cell of this subtle architectonic is arranged around convection points replying to the framework of a cosmos reinvented from the base to the summit, from the infinitely small to the infinitely large.

Following this beautiful example, everything here opens out into the most pleasing form of an intelligible phantasmagoria where the substance is never sacrified for the form, and where the form does not distort the function of the place and of its embellishments.

当客人一走进这个距离香榭丽舍大道不远、有如巢穴般神秘的地方，令人惊叹不已的是那些为酒吧空间带来谜样氛围的网状结构体，在此，时间已不存在。

这些仿佛要进行宗教仪式般的椅座像是隐藏在一艘淹没海底的雄伟船舰里的珊瑚珍宝盒。我们也可以说，这些具有有机形体的蜂房状结构体，犹如蚕茧般地排列相连，却无意撞见一个无法感知、不可预见、游移在两个无限之间的世界的孕育过程。当人们更仔细的观察这些椅子，会发现它们的椅座像一块紫色的集块岩，象征原始石海世界，也成为这个似乎在膨胀中的铝制结构的基础。接下来，看看椅子结构，这独特结构体的每个基本单位都依绕着对流点而排列，犹如一个重新再造的宇宙网络，从基点到最高点，从无限微小至无限浩大。

这里所呈现的一切形式都带着知性的魔幻与神奇，然而，空间底蕴绝不会为了迁就形式而牺牲，而形式的发展也绝不会违反或扭曲空间的功能性及其装饰美感。

设计师 Designer : Claudio Colucci
地点 Location : Tokyo, Japan
完工日期 Completion Date : 2007
摄影师 Photographer : Claudio Colucci Design

MOPH

莫非酒吧

The Moph space, originally designed in 2002, has been entirely redone, for sophisticated – even a little naughty! – adults. The global concept of this café-restaurant was developed by Claudio Colucci in consultation with the client. He ended up designing not only the space but also the plates and cups. Even a cake has been specially designed to make this place a total experience!

An emotive red has been selected for the ceiling and walls, while Colucci chose grey tables, based on Barcelona Style. The plain orange and translucent mesh curtains can slide in front of each other and adjust the balance between the inside and the outside. The heart-shaped "Petit Coeur" chairs designed by Colucci are made from red fabric and natural wood. Visitors feel embraced by a heart when they sit down in these sweet and poetic chairs.

Colucci likes to put a fun touch into each of his projects, and this "pop" taste helps to avoid the restaurant taking itself too seriously. It is a way to convey something to the client and for him to feel at ease and comfortable in a place he has just discovered.

创于2002年，由设计师克劳迪奥-寇鲁齐构思的莫非酒吧，于2007年重新由同一位设计师来对它进行全面的翻新，其所针对的是讲究雅致品位、甚至有点调皮叛逆的成人客户群。寇鲁齐不止为莫非酒吧设计了室内空间，也为它设计了家具与杯盘餐具，加上这里所提供的一种独创的糕点，带给客人一种全面而独特的体验！

设计师为天花板和墙面选择了一个充满激情的红色，而桌子则采用暖灰色，呈现出西班牙巴塞罗那的酒馆风格。以不透明和半透明布面交错排列的橙色窗帘可轻易滑动收放，调整厅内与厅外之间的平衡关系。由寇鲁齐特别设计，名为"小爱心"的心形座椅以红色织布与天然木材组合制成，当客人坐进这充满诗意又极为可爱的椅子时，感觉犹如被一颗心包裹着。

这个带有波普风格品位的设计，打破了一个高雅空间可能带来的严肃感，将趣味感融入设计的每个细节，让客人在发掘新空间的同时感觉轻松自在。

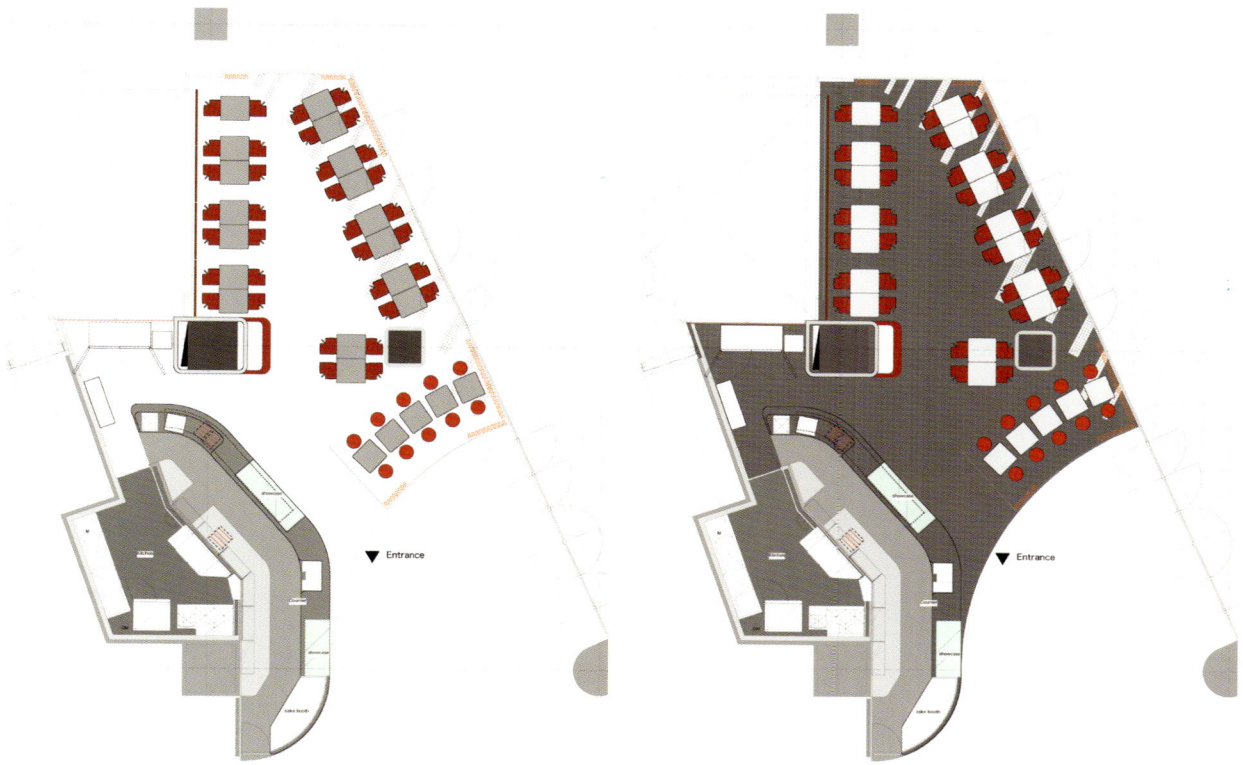

设计师 Designer : Christophe Pillet
地点 Location : Saint-Tropez, France
完工日期 Completion Date : 2008
摄影师 Photographer : Olivier Martin-Gambier

BAR DU PORT

海港酒吧

"A bright, luminous space, reflecting the port, the boats, absorbing the sun and the light…" was Christophe Pillet's inspiration for the redesign of the Bar du Port in Saint-Tropez. The Bar du Port is a family affair whose decoration dates back to 1963. When Christophe Pillet gave it its "first" new look in 2001, this rapidly placed the bar as one of the essential nightspots of Saint-Tropez. In 2008, when the owners were ready for a real change, Christophe Pillet decided to propose a whole new establishment.

After two months of intensive work, Le Bar du Port in its 2008 version has been entirely remodelled, based on a subtle play of mirrors, shadow and light. The latter two elements, which are so present in Saint-Tropez, were treated as key to the new decor. The floor is in slate-coloured ceramic tiles, which contrast with the white lacquered bar stools. Glass, a ceiling of smoked mirrors that doubles the height of the bar and gives it a freer space, walls in the white, roughened "Saint Hubert" stone and white furniture also create a dazzling ambiance in the Bar du Port, which is subtly transformed over the course of the day by three LED screens diffusing changing mood colours.

"一个清爽明亮的空间，反映出海港与船只，吸引了艳阳与光线……"，设计师克里斯多夫-皮耶替位在圣特罗佩市的海港酒吧重新塑形的灵感来自于此。海港酒吧多年以来都是家族经营，因此其室内装修从1963年留传下来，直至2001年。克里斯多夫-皮耶给予它"第一个"新面貌，使海港酒吧很快地成为当地不可错过的热门酒馆。2008年，在一种想改变、想更超越的欲望驱使下，设计师与酒吧主人共同决定再创新局面。

经过两个月密集施工后，海港酒吧2008年新版已完全改造出来。整个空间巧妙运用了镜子的反射效果、光与影的对比；在圣特罗佩，光与影两者都是此城处处可见的美景特色，因而也成为此次室内设计的重要关键元素。厅内地面是深灰色的缸瓷石板，与放置其上的白漆吧台高凳形成对比。天花板运用玻璃和烟镜，让人感觉整个餐厅的高度增倍，使空间感觉更自由舒畅；墙面采用的是"圣于伯尔"石，一种石面凿毛的白色石材；家具也一律白色，为海港酒吧创造出一种明亮耀眼的氛围，但这氛围也会随着墙上三面LED显示屏所映出变幻多彩的影像而不知不觉地转变。

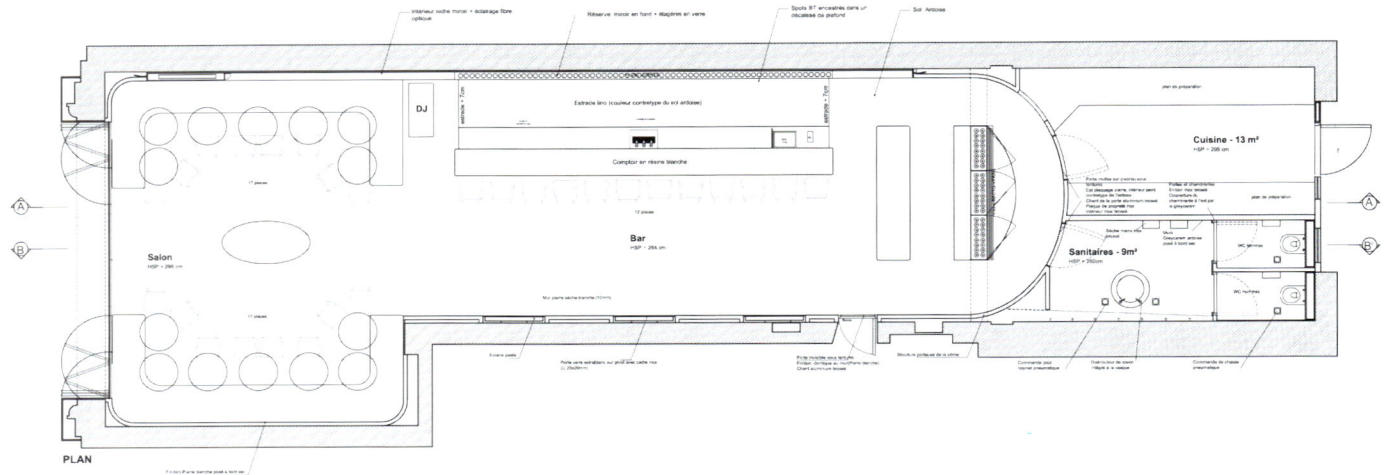

PLAN

设计师 Designer : Christophe Pillet
地点 Location : Paris, France
完工日期 Completion Date : 2008
摄影师 Photographer : Olivier Martin-Gambier

GUILO GUILO

枝鲁枝鲁餐厅

Imagined as a stage, upon which Chef Eiichi Eda Kuni performs his culinary arts, the brand new restaurant Guilo Guilo, designed by Christophe Pillet, has recently opened its doors in Montmartre. Echoing the famous novel "In Praise of Shadows" by Jun'ichiro Tanizaki, the designer conceived a space drawn with light and shadow, where simplicity and discretion are of the utmost importance.

In the middle of the main room, bathed in clear light that illuminates the dishes, is the chef's station, where the preparation of the cuisine takes place. An immense, metallic, rectangular block overhangs and dominates this space, serving technical requirements while also centring the view and the feeling of presence towards this "kitchen". Visual before it is visceral, Guilo Guilo immerses the diner in a complete experience: everything in its own time.

Diners sit on high stools designed by Christophe Pillet around the counter. Surrounded by the warm grey of slate and the sandblasted wood panelling tinted with a soft black that contrasts with the preparation space and is both enveloping and soothing, the audience-diners take part in a communion of both souls and senses, enjoying a cuisine that is drawn directly from the Japanese tradition. A small, red salon, comfortable in a pure design spirit, prolongs the space while offering a limited view through its opening.

由设计师克里斯多夫-皮耶构思的新餐厅——枝鲁枝鲁，在巴黎蒙马特山丘上开幕，整个餐厅被设计得犹如一个展演舞台。由日本名厨枝国荣一在此演出他的厨艺。设计师受日本文学大师谷崎润一郎的名著"阴翳礼赞"的影响，建构出一个呈现阴影与光线对比的空间，秉持单纯简约、含蓄、审慎的设计原则。

在令菜肴看起来更绝佳美味的大胆明快的光线下，主厨的展演空间如此坦率毫不犹豫地占据餐厅中心的位置，成为准备美馔的地盘。一个三大长方形金属体从天花板高高向下延伸，主导整个空间。如此设计不只是烹饪技术层面的需要，也引导客人视线投向这个"厨房"空间。枝鲁枝鲁邀请客人参与一段完整的美食之旅，先体验视觉震撼，再满足味蕾享受。

宾客围在台面四周，坐在由克里斯多夫-皮耶设计的高脚凳上，沉浸在椅凳的暖灰色与桌面板岩的深灰色色调之中，而四周墙面则以漆成柔和黑色的磨砂木板来装修，塑造出一种将人笼罩其中的轻松和谐氛围。宾客在此也成为观众，参与一场灵魂与感官的飨宴，品尝最传统道地的精致日本美食。餐厅内还有一个小厅房，艳红舒适，简洁且极具设计感，巧妙地延伸了餐厅的空间感。

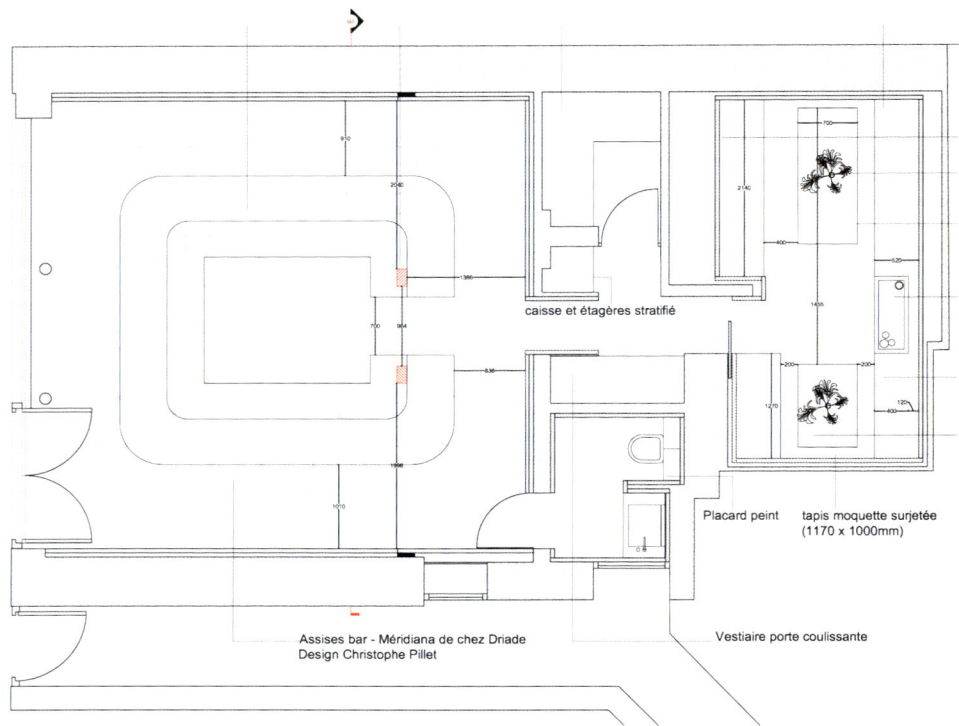

caisse et étagères stratifié

Placard peint   tapis moquette surjetée
(1170 x 1000mm)

Vestiaire porte coulissante

Assises bar - Méridiana de chez Driade
Design Christophe Pillet

设计师 Designer : Stéphane Maupin &
Nicolas Hugon
地点 Location : Paris, France
完工日期 Completion Date : 2003
摄影师 Photographer : Philippe Ruault

TOKYO EAT

东京食餐厅

The restaurant has a simple concept that accompanies the artistic experimentation of the site (the Palais de Tokyo contemporary art space in Paris). The kitchen is the centre of the action and is visible to everyone. The project divulges how the dishes are made by exposing the cooks at work.

A large, translucent box hangs from the ceiling. The sun's rays enter through its South-facing window, cross the space and cast long shadows when they encounter the diners, thus playing a part in the animation of the space. This box contains WCs from all over the world. These translucent toilets play on the relationship between seeing and being seen, inherent in the design of fashionable restaurants. This collection of world toilets actually forms part of the museum's collection, exciting curiosity for the art itself, and includes a German WC, a Japanese WC with its self-cleaning bowl, a large American WC, an Indian WC, an Italian urinal and a French child's WC... This international collection is accompanied by a rudimentary hand-washing system assembled from industrial protection materials for gas workers (a hand-washing shower and eye-bath). A lexicon of use is displayed in the form of generic ideograms.

As for the lamps, they can do just about everything. Suspended in the space, they can be moved the simple touch of a hand. Their position on a grid is controlled by a central lighting island. The changing altimetry expresses the different moods of the space. They look like a series of ciboriums lighting up a church. A tall nave for the ritual of the meal surrounded by contemporary coloured glass windows (by the artist Beat Strülli). The ideal height of these 1.2 metre diameter lamps is 80 centimetres from the table, where a striking contrast between the hanging (+ 7.00) and the point of the light's diffusion of takes place. The light remains limited in proximity, avoiding dispersal across the whole space. In order to respect the correct sound ambiance, speakers are hidden in the lamps, bringing the consumer into an even more personalized atmosphere. The lamp is everything at the same time, screen, entertainment, light source, sound source, decoration...

这个位于东京宫当代美术馆内的餐厅，其室内设计配合美术馆的艺术实验风格，呈现出极其简洁的特色。厨房被设计成是整个餐厅最引人注目的活动焦点，完全展露在众人眼下，透过师傅们的厨艺展现，来揭开美食的秘密。

一个半透明的长盒状服务性空间从餐厅天花板悬置下来，来自南面的光线照亮了整个餐厅，也投射到这个悬高的空间，使得楼下的客人能在半透明的隔板上瞥见这个盒状空间使用者的身影，成为有趣的一幕。这个空间装设了各式各样的马桶，而这些半透明厕所制造出的一种近乎偷窥与被偷窥的关系，也正呼应着当代社交餐厅的设计概念。为了吸引众人的好奇心，这些来自世界各地的马桶也成为美术馆的收藏品之一，包括德国马桶、日本马桶及其自动清洁马桶圈、特大号的美式马桶、印度式马桶、意大利式小便壶、法式儿童专用小马桶等等。这些包罗万象的全球性收藏品还伴随着一排形式原始简朴却极为特殊的洗手槽，由煤气工业的保护清洁装置来组制而成，带有莲蓬头式的出水口和洗眼装置。这里的各种器具使用说明以特别设计的图像形式标示出来。

餐厅中还悬置了一系列多功能灯具，让人能随心所欲地用手调整变动。吊灯的装置方位配合餐桌位置而形成一个个照明区域，可变动的悬吊高度也显示了空间的宽广与变化的弹性。这些灯饰看起来极像一个个的圣体盒，让整个餐厅犹如一座教堂。这种犹如身在殿堂内的眩然之感，强化了在环绕着当代彩绘玻璃（艺术家比特-史楚利的作品）的空间中进餐的仪式性。这些直径长达1.2米的球形灯具的理想悬吊高度是距离桌面80厘米高的地方：吊灯挂钩和光线散发点之间产生明显的光影对比，而且光线能够聚集在客人周围却不会扩散到整个空间。此外，为了创造适当的声效，音响喇叭也被装置在灯具上，使客人沉浸在一种属于个人的亲密氛围之中。此灯具发挥着多样功能，是光线的来源，是音乐的来源，是屏风　是装饰品，也是趣味的焦点……

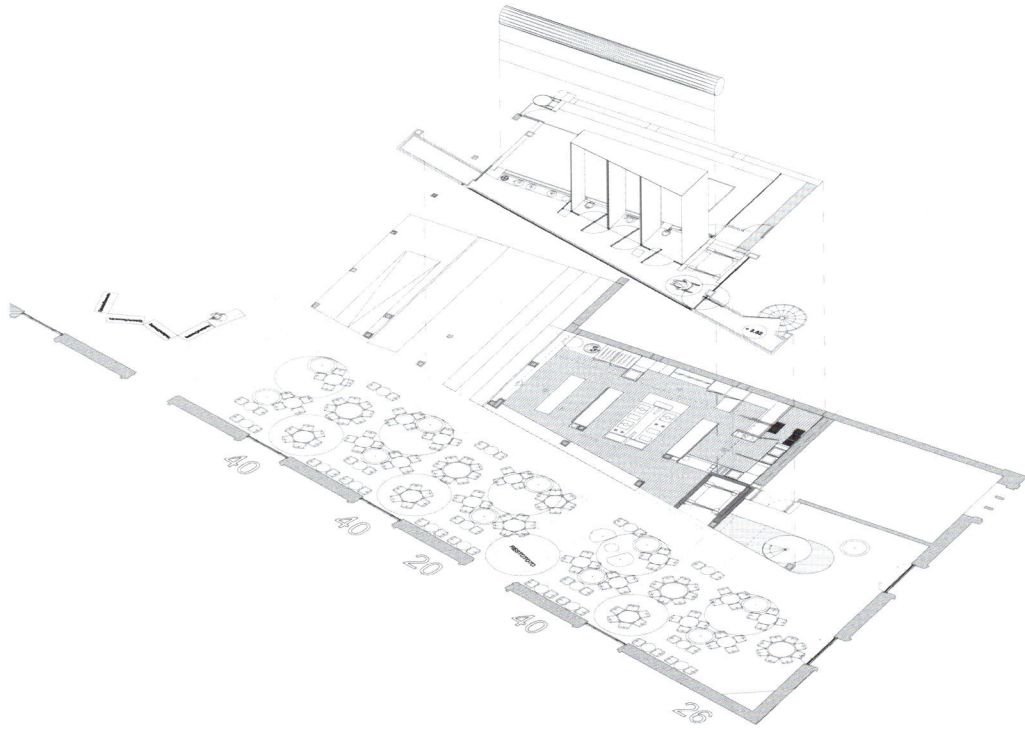

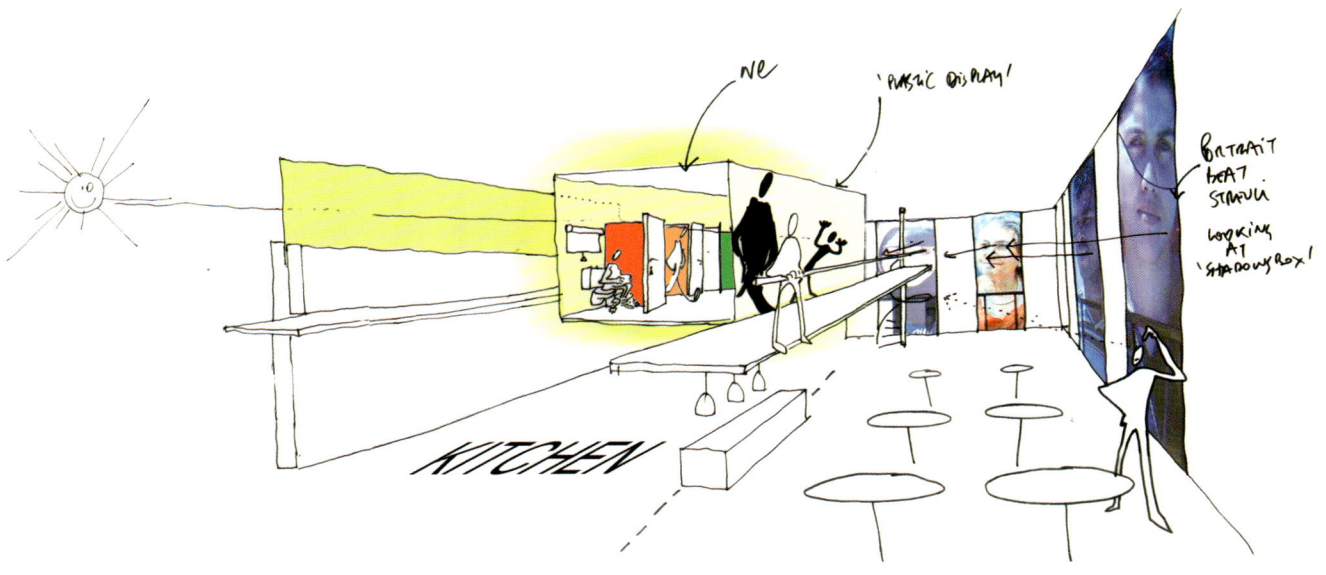

设计师 Designer : SCP d'architectes,
Denis Boyer-Gibaud,
François Percheron, Antoine Assus
地点 Location : Paris, France
完工日期 Completion Date : 2008
摄影师 Photographer : Didier Boy de la Tour

BABOTO

八波头餐厅

The multi-faceted Baboto has opened its doors in rue de la Ferronnerie, Paris. The restoration of this magnificent building, dating from 1669 and classed as a historic monument, took two years. It has been a restaurant since 1860: the Maison Rouge, as it was then, in the beating heart of Paris, between the Faubourg St Honoré, Les Halles and the Montorgueil quarter.

Why Baboto? Because the place refers to the South and to Montpellier in particular. An old entrance door from this city "where the sun never sets" is called the Babote. In the Occitan language "baboto" is also a name given to imaginary people, fairies or little monsters. From the outset, Daniel Alauze wanted to create a place with the atmosphere of the South, a place for a modular lifestyle, where one wants to spend time in a vibrant neighbourhood. "A chic and relaxed spot (...) with a designer decor, but design for living, not just for looking at".

The complete renovation by the Montpellier architect-designer Denis Boyer-Gibaud combines the original stone with glass and exotic wood (ipe). The place is very minimalist, and the design appears in the smallest details. There are several living spaces where one feels at home. The materials are warm and the colours soothing: red and deep khaki. The bar, which is also very conceptual, is "dressed" in red Dacryl, a high-end synthetic glass giving plenty of cachet to the whole look.

八波头这个面貌多元的餐饮空间不久前在巴黎市费宏纳里路全新开幕。餐厅位在一栋建于1669年、已被列为国家级古迹的宏伟建筑中，先前这栋楼的整修工程长达两年之久。自1860年以来，此地成为著名的"红屋"餐厅。地理位置优越，介于弗埠-圣-欧诺黑名店街、市集商场和蒙陀戈伊街区之间，位处巴黎心脏地带。

为何取名为"八波头"？因为此餐厅要呈现的是法国南部的美食与风情，尤其令人联想到蒙彼利埃市，在这个"日不落"之城有座老城门就称作"八波头"。在奥克语中，"八波头"也是一些传奇小人物、仙女或鬼灵精的名字。丹尼尔-阿楼兹在方案初始即意欲创造出一个拥有南方风韵的空间，一个变化多端的生活空间，他希望在一个充满活力的城区中，营造出一个令人想长时间逗留的地方。"一个轻松舒适又时尚摩登的餐厅……，并且有着极具设计感的装修，但不是仅供观看的设计，是让人能体会生活、令人活在其中的设计。"

整个装修工程由来自蒙彼利埃的建筑设计师德尼-波瓦耶-吉伯负责，他结合原产地石材与玻璃，并搭配使用南美洲的重蚁木。餐厅设计极其简洁，直至最微小的细节都经过缜密的设计，使得厅内拥有多个令客人感觉十分舒适的生活空间。装修所运用的材质都很温馨，搭配着平和稳重的色彩；红色及深卡其色。设计感十足的酒吧，采用红色的达克利乐材质，这是一种高级的透明合成塑料，为餐厅整体带来高雅时尚的感觉。

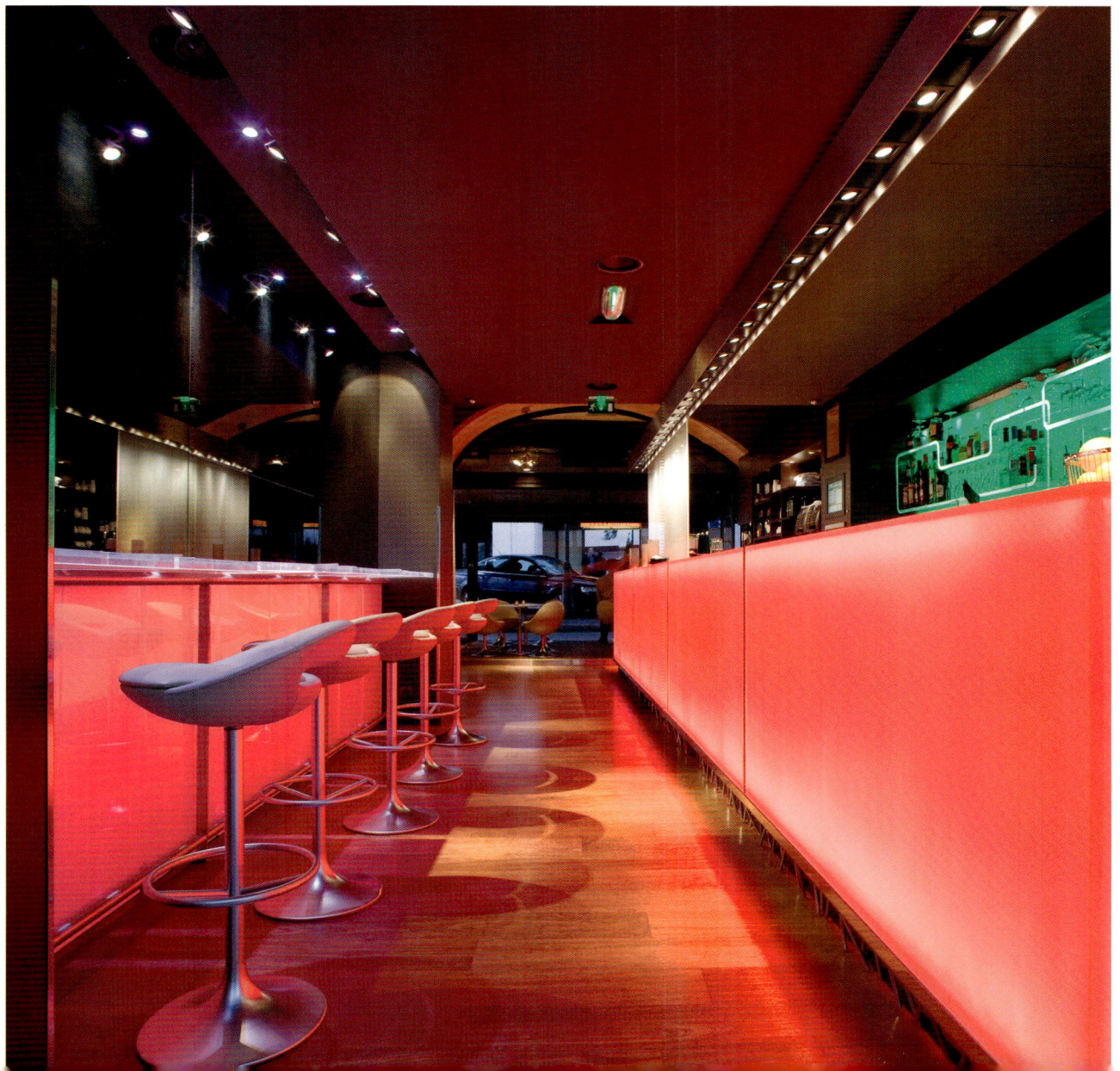

设计师 Designer : Jean-Pierre Mallet,
Agence Basic
地点 Location : Paris, France
完工日期 Completion Date : 2006
摄影师 Photographer : Wijane Noree

LE TARMAC

塔马克餐厅

The transformation of this old brasserie prioritizes the notion of space in order to make the most of the existing volumes. The 19th-century painted glass ceiling has been preserved and made a part of the new modernity of the place. The architecture is by Jean-Pierre Mallet of the Basic agency, and the graphic design by the Morellongiralt studio in Barcelona.

Actors and spectators come together on the same stage: a large bar, kitchen, "bistro" space and dining room are blended in a single open space. The interior structure is composed mainly of wenge wood, stainless steel and concrete, and softly lit. Thin, painted wooden slats for the Venetian blinds, Corian and pure white laminate for the tables, intense red imitation leather for the banquettes, orange and white chairs with stainless steel feet, enormous lampshades like upside-down flower pots in translucent white, a fresco of over-tinted, black-and-white or projected photographs, evocative gobos, grey plasterwork and calligraphy effects make up a decor that is both classic and innovative.

On the exterior, the dark grey facade, its large black awning with red and metallic grey motifs and a permanent tree-filled terrace that is heated in winter naturally beckon the traveller to stop on the "Tarmac".

这间古老餐馆的全新室内装修将设计重点强调在空间的概念，使现有的空间体量呈现更宽阔的感觉。19世纪的彩色玻璃天花板被完整地保留下来，并且融入了极具现代感的新场所当中。餐厅的建筑设计由法国的让皮埃尔-马雷（基础事务所）负责，而美术设计则由巴塞隆纳的摩瑞龙吉拉工作室来进行。

在此餐厅中，演员和观众全集聚在同一个舞台上：大型吧台、厨房、酒吧间和用餐区全融合在同一个开放式的空间里。在柔和的灯光下，厅内空间主要由红木、不锈钢及混凝土装修而成。美式的遮阳帘幕有着上漆的细木条，餐桌采用可丽耐和白色层压面料，长软垫椅是超红的斯凯彩膜质材，橙色及白色的椅子配上不锈钢制的椅脚，一盏半透明的白色大吊灯形似倒过来的花盆，一系列色彩极度鲜艳、或黑白、或甚至投影呈现的照片，唤醒记忆的图案片，粉饰灰泥和书写字体的墙面，整体组合成既古典又革新的装饰设计。

餐厅外观是黑灰色墙面，搭配着带有红色和铁灰色图案的黑色遮阳帘幕，这里还有一个四季植物繁盛、冬天设有暖炉的露天座，以舒适幽美的环境吸引游客在塔马克餐厅停下来歇息。

设计师 Designer : L'Agence (Interior Design)
地点 Location : Paris, France
完工日期 Completion Date : 2008
摄影师 Photographer : L'Agence (Interior Design)

SHAKE EAT

雪克食餐厅

The restaurant Shake Eat was envisaged as an urban kitchen garden, a place where one comes to fill up with fresh and healthy food with a fit and sporty dimension. First of all the architects wanted to open up the area and make it luminous by making space around the plate glass window and using light colours. In this way they laid down the basis for a feel-good space. Next, attention was paid to the lighting in order to bring intimacy to these open spaces, using bell-shaped shades hanging from the ceiling of the ground floor, and standard lamps on the first floor for example.

Two vertical kitchen gardens were created on the ground floor and the first floor, in order to integrate the presence of plants, which was indispensable to the concept of this fit-out. The colours raspberry and aniseed allowed them to add a fresh and sporty touch to the restaurant and have also been used in the visual identity and the signature of the place. To enliven and give a rhythm to the interior, touches of stylised nature have been used in different forms: stickers on the wall, a bookcase with a plant motif, a luminous setting for plants. All these touches help in the creation of interior landscapes and, the creators hoped, give a general impression of being in the presence of natural harmony.

The ecological impact of the restaurant was a preoccupation throughout the project in the choice of materials and their exploitation: wood on the floor and on the tables, stoneware and ceramics, low energy light bulbs, no air conditioning but natural ventilation, etc.

雪克食餐厅被视为一座都市里的菜园来设计，提供一个人们可前来享用新鲜健康美食的地方，一个令人精神振奋、充满运动活力的地方。室内设计的首要任务是先将空间释放开来，运用大面的落地窗及浅色主调让空间明亮鲜活起来，以此建造出一个令人感觉舒适的空间基本架构。接着再精心设置灯光，为这些开敞的空间带来温馨亲密的氛围，例如，地面楼天花板的撞钟形吊灯，以及一楼的立灯。

为了将设计概念中不可或缺的元素——天然植物——融入此装修空间中，地面一楼和二楼各创置了一个直立式园圃。覆盆子色和茴香色为餐厅增添了精神与活力，也被运用在整体的视觉设计上，成为餐厅的独特标记。为了让厅内更活泼更有节奏感，一些风格化的自然界元素以不同形式呈现，如：墙上贴纸、植物形状的书架以及种植绿草的亮光框等。这些巧妙的装饰一起创造出一幅美丽的室内风景，且如同诸位创立人的期望：使餐厅的整体印象给人一种犹如身处一片自然和谐之中的感觉。

餐厅在整个装修过程中都十分坚持环保理念，注重材质的选择及其使用方法：地板采用木材，餐桌用陶土和陶瓷，灯光用省电灯泡，不装置冷气空调而采用自然通风等。

设计师 Designer : Ora-Ïto
地点 Location : Paris, France
完工日期 Completion Date : 2008
摄影师 Photographer : Olivier Sochard,
Massimo Pessina

KAISEKI BENTO

怀石便当餐厅

In this European flagship, Toyota and the designer Ora-Ïto wanted to complete the visitor's experience with an invitation to eat in a welcoming space that matched the philosophy of the place. To achieve this, Ora-Ïto above all wanted the bar's interior architecture to be modular. The space, with its curvilinear bar, is set up for the preparation of made-to-order bentos and designed to be staffed by two to three people. The shape and minimalist feel of the bar are identical to those of Le Rendez-Vous Toyota. The back wall features a germination space created by Elisabeth and Hisayuki Takeuchi, firmly anchoring the concept of healthy food in the restaurant's interior decoration.

In the dining room, the designer opted for a mix of tables and seats for up to 24 guests. Ora-Ïto treated the entire space much as a stage designer would. The design of the seats evokes pieces of ripe fruit. The cushions made of soft materials in a variety of colours, contrast sharply with the immaculate white Corian structures. The round tables were designed so that four of the chairs would slide right under them easily. Once the chairs are pushed in, each grouping forms a compact, stackable-looking cylinder.

When it is closed, the bar is transformed into a lighting display. With its backlit curtains, it looks like a huge, horizontal wall light that casts a glow over the entire area.

日本丰田汽车与法国设计师欧哈-伊都希望在这家位于巴黎的欧洲旗舰店里加入一个与展示空间具有相同温馨迎人气氛的餐饮空间，让来参观新车展示的人们能够在此用餐。设计师为此而进行的首要构思，是将吧台的建筑结构设计成一个可灵活变换的模柜式设备。这个具有圆角线条、可容纳二至三个服务人员的吧台空间可以依照客户的点餐情况或盒餐的制作需求来调整变动，其造型设计极其简约纯净，与丰田之约展示馆的整体空间风格产生和谐一致的搭配。厅内最里边的一道墙面上饰有植物萌芽般的图案，是伊丽莎白和日本名厨久之竹的构思，呈现出此餐厅所倡导的对身体有益的健康美食原则。

在用餐区内，欧哈-伊都设计了可容纳24位宾客的餐桌椅组合，他把整体家具设计想象成可随时装卸的舞台装置。这里的桌椅组合仿佛是一颗颗令人想咬一口的鲜美水果，座椅的色彩与灯饰相呼应，而其柔软的材质正好与座椅结构那洁白无比的可丽耐材质产生强烈对比。餐桌则被设计成能让四张配套的座椅很容易地靠收进去，而且一旦座椅全组收合到餐桌底下，整体便成为一座圆形的台子，可轻易地与其他桌椅组合叠置在一起。

在餐厅的开放时间之外，吧台摇身一变，成为一个照明装置，掩上一层布，从里头打灯反照。饰了帘幕的明亮吧台看似一盏垂直型的大壁灯，使整个空间在非用餐时间也显得美轮美奂！

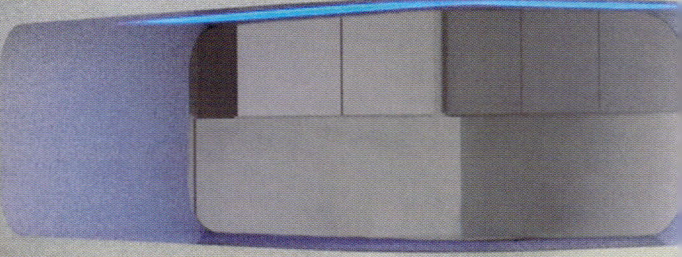

eat healthy be my b

yaourt glacé 0% de matières grasses, 100% fruits

alliez gourmandise et sérénité...

设计师 Designer : Claudio Colucci
地点 Location : Paris, France
完工日期 Completion Date : 2008
摄影师 Photographer : Gilles Toledano

MYBERRY

吾之莓冰淇淋馆

A space led by the white, with curved and elegant shapes, which is the illustration of the restoration concept: "0% fat yoghourt". This space was designed with the same movement once need to serve a cup (cutting) of ice-cold yoghourt. To say it in another way, the gesture of the maitre glacier was extrapolated to synthesize it in an architectural space.

The outside facade clearly reveals this idea of Claudio Colucci's concept and already suggests the tasting. The bar counter is not more than the logical continuity of this white, elegant, dynamic and curvy material. It is a place with no free effects, with only slight chromatic touches, a variation of refreshing colors just like the products which will come to complete the yoghourt. Colors or materials appear behind the white, as if it had been removed with ingenuity and gourmandise, to draw the space and create features: niches of colors and materials behind a white volume.

The graphic identity is more malicious, even though turned to the same search for graphic minimalism: a simple typography illustrating the idea of gourmandise and pleasure thanks to the drawing of a face in the happy smile.

这是一个以白色为主调、以圆弧线条及优雅形体来设计的空间，符合"零脂肪优格"的新餐饮概念的形象。设计师的灵感来自挖一杯优格冰淇淋的动作。也就是说，设计师将挖取冰淇淋的动作转换成设计此建筑空间的构想线条。

店面的外观设计已明显地揭露设计师的这个构思，而且呈现出产品的诱人美味。冰淇淋柜及吧台理所当然地延续了这白色的质感及优雅又有活力的曲线。整体空间的形体设计具有简约纯净的格调，唯一运用到的色彩点缀便是呼应那些添加在优格上的加味品的新鲜色彩。这些色彩和各种材质出现在整体白色背景下，仿佛部分白色被贪嘴馋客巧妙地取走，以便建构一些小空间画并赋予某些功用性。因此，在白色体量当中时而出现一些色彩亮丽、具有不同材质的凹巢空间。

商店的识别标志虽然也是朝向简约风格，却具有较为活泼调皮的精神：一个仿佛拥有光彩笑容的脸型构想，使得极为单纯的字样也能表达出品尝优格冰淇淋的美味可口和愉悦享受。

myberry®

Glace au yaourt 0% de matières grasses, 100% fruits

eat healthy be my berry

yaourt glacé 0% de matières grasses, 100% fruits

alliez gourmandise et sérénité...

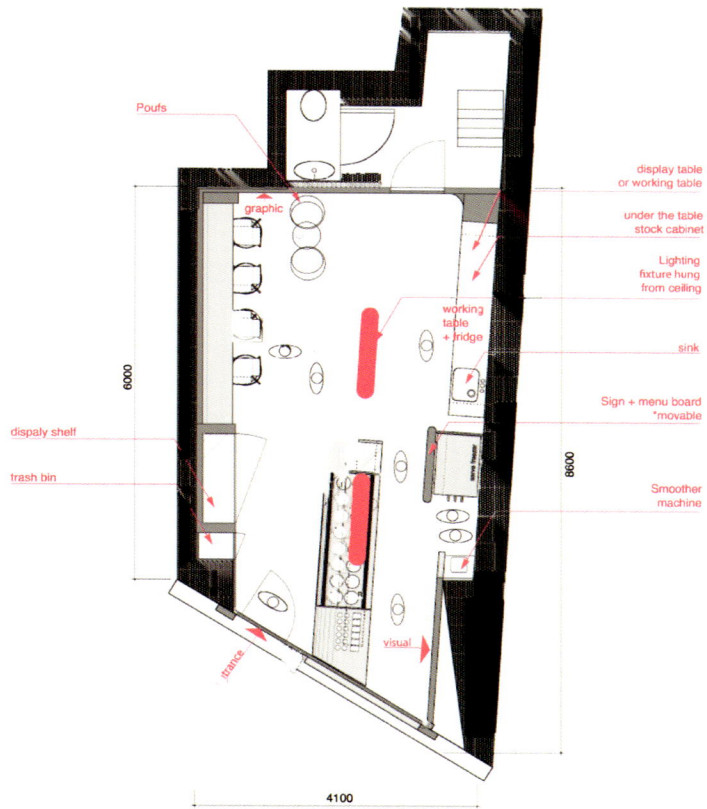

Poufs

graphic

display table
or working table

under the table
stock cabinet

Lighting
fixture hung
from ceiling

working
table
+ fridge

sink

Sign + menu board
*movable

display shelf

trash bin

Smoother
machine

6000

8800

4100

entrance

visual

设计师 Designer : Idoine
地点 Location : Paris, France
完工日期 Completion Date : 2005
摄影师 Photographer : Henri Perrot

EATME

食我餐厅

In the Montorgueil quarter of Paris, the restaurant Eat Me, designed and fitted out by the design agency Idoine, offers a real moment of relaxation and pleasure. To break with the uncomfortable lunchtime habit of sandwiches hastily consumed in noisy, narrow spaces, the agency set itself the challenge of reconciling a balanced and delicious cuisine with the pleasure and comfort of a design environment... all with a feeling of lightness!

Idoine was able to give the place its heart and originality by creating a serene and calm atmosphere on two levels. A sober and poetic design ethos is embellished by site-specific indirect lighting. The chosen colours are in tonalities of putty and taupe, enlivened with touches of coral and citrus green.

An aromatic herb garden grows on the large, suspended communal table on the ground floor, and custom-made swing seats seen through the window invite one to picnic. An alcove with a cosy ambiance is found on the mezzanine, from which the François Azambourg light forms a magnificent, luminous and mobile counterbalance to the communal table. A delicious atmosphere providing the perfect setting for a mealtime break where one can enjoy eating with a feeling of weightlessness!

这家名为"食我"的餐厅位于巴黎蒙陀戈伊街区，由异端建筑设计公司一手包办进行设计及装修，旨在提供食客一段全然放松享受的时刻。很多人的午餐都是在一些吵闹拥挤的地方胡乱吞食几口三明治打发掉，为了排除这种极不舒适的饮食习惯，食我餐厅要把主要设计目标定在：如何提供客人在一个富含设计品位的舒适空间里享受一顿既营养均衡又美味的餐点，而且整体呈现出一种轻盈之感！

异端设计公司为这个餐饮空间创造出了属于它自己的灵魂和个性，在两层楼的空间中营造出静逸安详的气氛。为此，设计师采用了简洁又饶富诗意的线条，再配上特殊的间接照明式灯光，色彩上则运用灰黄色和暗黑色调，并以珊瑚红和鲜绿色点缀以提高整体空间的明亮度。

餐厅一楼以悬吊方式设置了一张长方形的大主桌，桌上长了一园子似的芳香植物；为靠窗位置特意设计的荡秋千，令人联想到郊外野餐的景象。夹层上构建了一个气氛舒适的凹室厅房：一盏由弗朗斯瓦-阿赞堡设计的吊灯从夹层上方天花板悬挂下来，成为与大主桌上下相对的一个明亮又具动感的平衡体。餐厅中的愉悦氛围完美地表现出一个可以让人飘飘然地品尝美食的休憩时光！

baramaki

设计师 Designer : Anégil
地点 Location : Paris, France
完工日期 Completion Date : 2006
摄影师 Photographer : Xavier Béjot

BARAMAKI

巴哈马吉寿司餐厅

The Baramaki stands out in the heart of Printemps Haussmann, the definitive home of Parisian fashion. With an original concept, this sushi bar takes its inspiration from the circular plan of one of the domes.

The crazy and organic route of the "sushi-train" thus accentuates the layout of the premises. The conveyor belt in stainless steel serves the "tentacles" of the lacquered grey tables, giving on one side a face-to-face view of the sushi-man, on the other a new take on communal tables. The floor defines these two spaces, passing from a dark oak parquet to a thick red lacquer. Chairs specially designed in acid colours, with bases finely lacquered in black, punctuate the space.

To the side, an oversized bento box symbolises the bar, echoing the striking red of the floor. The space behind is covered in shining black laminate reflecting the atmosphere all around. The spirit of the place is symbolised by a large luminous screen on the periphery showing images that evoke the rhythms of the seasons. These refined landscapes of Japanese cherry trees are dissected by "reflective" curtains.

Looking up, one discovers an indirect lighting that draws the outline of the circular plan, as well as a graphic representation of a flower in profile, enhanced by three huge coloured lamp shades. The contrast of materials, colours and degrees of brilliance reinforces the "Shibu-esque" world created here.

巴哈马吉寿司餐厅开设在奥斯曼大街的春天百货里面，位于绝不可错过的巴黎时尚区内。这家寿司餐厅的设计十分独特，在百货公司著名的圆顶建筑里的其中一个圆形空间中。

运送寿司的回转火车的路线是偏离轴心的非几何形设计，因此更强化了餐厅内部的空间格局。不锈钢输送装置犹如触手形式服务着一旁的灰色亮漆台桌，形成客人与师傅面对面的座位布局，也制造出一种主宾桌形式般的亲切氛围。地面的不同处理方式界定出了两个空间，一边是深色橡木地板，另一边是红色厚漆地板。椅子设计特别运用带有微酸感觉的水果色彩，且椅脚也精巧地上了黑漆，为整个空间缀上节奏感。

在一旁，一个特大号便当盒形状的吧台也采用跟地面一样的艳红色。吧台后面装潢上亮黑色的层压材质，反映出四周氛围。餐厅的环形周围装置了一排明亮的荧幕，映出四季变化的韵律，也象征了餐厅的精神所在。在这些日本樱树极简极美的景色前，等距设置了几条隐约透光的布帘，从天花垂吊了下来。

抬头向上一望，天花板上装置的间接照明灯光与圆形空间互为呼应，三盏尺寸惊人的彩色灯罩烘托出呈现花朵切面的图案。材质、色彩和亮度的对比都一再强化出此餐厅的前卫日本气息。

À bientôt

Venez le soir de
votre anniversaire,
le gâteau c'est cadeau*

Demandez
notre offre traiteur,
pour vos petites et
grandes occasions

Nous vous accueillons
tous les dimanches
et les jours fériés

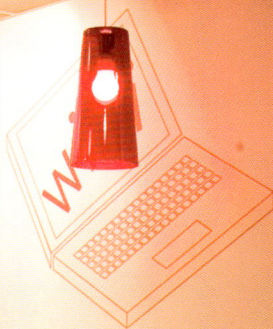

*Voir conditions en restaurant

glaces

Wifi

设计师 Designer : Dragon Rouge
地点 Location : several sites in France
完工日期 Completion Date : 2007
摄影师 Photographer : Arnauld Duboys-Fresney

FLUNCH

富兰奇餐厅

For this leading affordable restaurant chain, a new contemporary concept has taken root. Dragon Rouge Archi created a building with extremely refined contours and a highly modern profile, giving a feeling of being open to the outside, and adopting an appropriate, restrained tone perfectly in line with the corporate values defended by the food chain.

The self-service ritual here is given a highly flattering setting to emphasize the quality of the welcome and respect for the customer. The dining area has been carefully designed to satisfy the various needs of different types of clientele: a quick snack, or a shared moment of conviviality; a meal eaten alone or in a twosome, as well as meals eaten at different times of the day. Accordingly, the restaurant provides large dinner tables for meals with family and friends, "snack" tables designed for quick bites to eat, along with tea-room and fast-food areas forming a bridge between the restaurant and the take-away outlet.

The atmosphere in the dining area is serene and up to date: beige walls decorated with pictures of fruit and vegetables in matching tones, a taupe-coloured ceiling and wall-to-wall carpeting chosen to create a soothing acoustic environment, a subdued lighting scheme, furniture dressed in warm colours (light beech and wenge) set off by splashes of colour on the service elements of the furniture (providing coffee, water, sauce and dressings, etc.).

Lastly, the entire restaurant was designed to ensure that children are actively involved in their meals: a children's dining area has been created with furniture and buffet tables of the appropriate height along with a signage system specifically designed to help guide them, in a clear and playful manner, in the choice of their meals.

富兰奇这个连锁餐饮店委托红龙公司建筑部门为它提出一个具有当代感的空间概念：一栋线条极其现代，形式极其简洁的餐厅。建筑于是诞生，整体空间是向外开放的格局，设计格调十分精准朴实，十分符合富兰奇品牌的价值观。

在室内设计方面，设计师将自助餐饮的形式特色以一种具有高质量的仪式方式呈现，着重于提供客人良好的接待与尊重。一切细节都被仔细考量过，以满足不同客户层的需求，无论是快餐或亲朋好友聚餐，以及在一天中不同时刻的用餐，都能提供最佳的使用空间。因此，厅内设置了宽阔的大桌、适合享受随手拿取简餐点心的小吃台，甚至也有品茶的厅房。

厅内气氛十分轻松且极具现代感：淡褐色的墙上饰有同色调的蔬果图案、深棕色的音效天花板和能够消音使厅内不致于过度嘈杂的地毯、温和的照明气氛、暖色调的家具（采用浅色山毛榉木和红木制家具），还配上些许色彩鲜明、用来整合附属服务（提供咖啡、水和酱料等）的家具。

除此之外，整个餐厅还专为孩童构思了属于他们的空间，让他们能够自由自主地取用餐点，设计符合小孩高度的餐台餐桌椅，加上一套既清晰又充满趣味的标志来引导他们选用自己的食物。

La table de l'info

La table de l'info

La table de l'info

La table de l'info

Merci

Merci

Merci

Buffet enfants

Les grillades

Les g

les services Flunch
glaçon
eau
micro-ondes

FLUNCH**TV** FLUNCH**TV** FLUNCHT

4 ESPACES POUR VOUS
AREAS FOR YOU

2 **ice bäar**
BAR-LOUNGE
à partir de 19 heures ;
accès supplémentaire
au 41ter rue Moufert

1 **GOLD ROOM**
SALON DE THÉ
ICE CREAM RESTAURANT

0 **THE COLLECTION**
VENTE À EMPORTER / TAKE AWAY
Espressamente illy

-1 **DAILY CAFÉ**
SALON DE THÉ - CAFÉ

SHANGHAI.BERLIN.NEW-YORK.TOKYO.HONG-KONG.LON

设计师 Designer : Saguez & Partners
地点 Location : Paris, France
完工日期 Completion Date : 2007
摄影师 Photographer : Olivier Seignette,
Mickaël Lafontan

HÄAGEN DAZS

哈根达斯冰淇淋馆

Designing a colour palette that would systematically identify the place was the question addressed to the agency charged with designing this vast space. The response came in the form of a recipe: starting with the key ingredient of the brand, cream (the product) became a colour, to which was added Häagen Dazs burgundy with a sprinkling of gold.

The contemporary furniture of the terrace, parasols, tables and chairs, represents a visible change. The dominant tones of oatmeal, the famous "burgundy" and gold highlights allow one to identify the place with the brand. The Daily Café is a cosy and welcoming place, simply laid out around a large communal table, with a comfortable lounge corner or a counter at which to enjoy a coffee at any time of day facing a large plasma screen.

The Collection ice cream bar has a dark floor in ceramic tiles, furniture in brushed steel and transparent glass, and cupboards built into the walls. A white ceiling dominates this area and gives it coherence. A golden staircase at the back of the room creates a strong focus and attracts the consumer to investigate the levels above. The Gold Room is a tea salon organised around a dome giving onto the ice cream bar. White, gold and outmeal colourways are found on the furniture, the walls and the mirrors. A daring design puts together contemporary tables in white Corian and golden chairs. A floor in dark wood brings the feeling of materials and comfort. The whole room is treated with subdued light.

Finally, the ambiance of the Ice Bäar breaks with that of the other levels: a chocolate coloured wallpaper uses the monogram of the Häagen Dazs packaging. Seating squares behind magenta coloured glass serve as VIP areas and give rhythm to the space. A concrete floor shot with burgundy sequins and a bar in brushed steel complete the setting.

对建筑师事务所来说，要装修一个如此庞大的店堂，方案的重点在于如何构思出一套色彩系统，让人们能够马上很清晰地识别各个空间。构思后获得的答案仿佛为空间设计了一道食谱：从品牌的主导食材（奶油）出发，因此奶油色成为重要色彩，加上哈根达斯那带点紫调的暗红色，然后再处处撒上亮金色！

露天座的时尚风格家具：大阳伞和餐桌椅，很明显地表达了一种品牌风格的转变。米灰色主调配上著名的哈根达斯暗红色，加上金色点缀，让人一眼即能识别出是此品牌的店面。"每日咖啡"是个很温馨舒适的空间，整个区域呈现简洁的空间规划：一张大主宾桌，加上舒适的沙发区和高台座位区，面对着大液晶荧幕，可随时享受一杯咖啡。

"冰淇淋吧"有灰调的缸瓷地板、刷钢材质以及透明玻璃制成的家具和壁置橱柜。白色天花板主导这个空间并使一切看起来更协调一致。厅内最里边有座金色楼梯，其亮丽光彩吸引客人前往楼上。"金厅"是饮茶沙龙区，呈圆弧形楼中楼空间格局，往下望去便是"冰淇淋吧"。区内的家具、墙面和墙上的镜子皆采用白色、金色及米灰色互相搭配。设计师以大胆的手法将白色可丽耐材质制成极具当代感的餐桌和金色复古风格的椅子搭配在一起。暗调木质地板为厅内带来特殊质感和舒适感，而整体空间则沉浸在一片柔和的灯光下。

"冰淇淋俱乐部"的气氛则跟其他楼层截然不同：巧克力色的壁纸上印有哈根达斯标准字的交织图案。紫色玻璃建造出来的一系列方盒子作为贵宾室之用，并且为整体空间带来韵感。混凝土地面灌撒了哈根达斯的标准暗红色亮片。这里还有一座刷钢材质的吧台，让整体空间设备更臻完善。

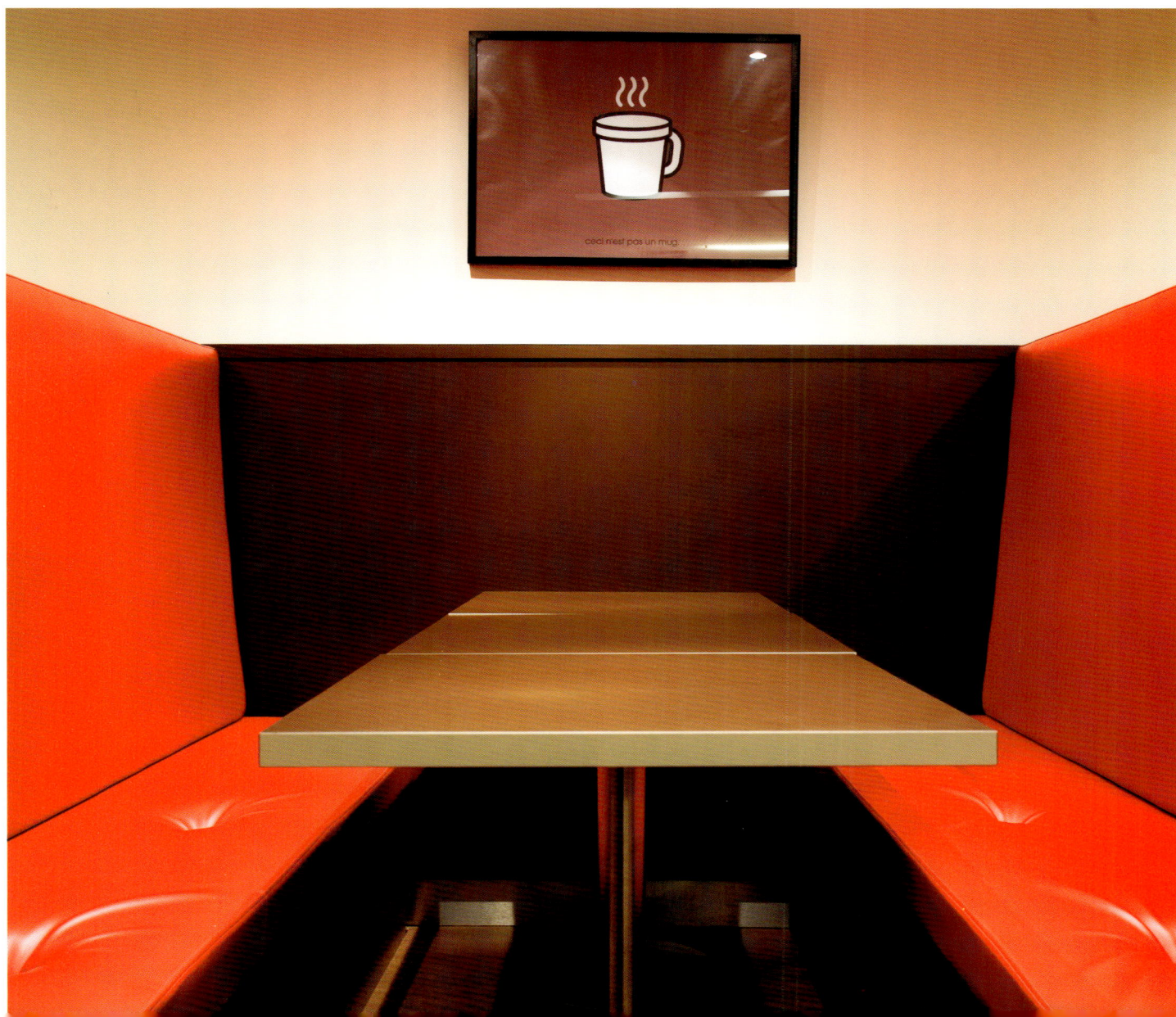

ceci n'est pas un mug

设计师 Designer : Christian Ghion
地点 Location : Tokyo, Japan
完工日期 Completion Date : 2007
摄影师 Photographer : Christian Ghion

PG CAFE

PG咖啡馆

Continuing his collaboration with Pierre Gagnaire and faithful to their common understanding, Christian Ghion has created a new space in Tokyo, which was opened in Autumn 2007: the "PG Café" (Pierre Gagnaire Café).

The "PG Café" is a Japanese-style corner café where snacks, soups, salads and ice creams are served. The café enjoys an exceptional location mid-way between Tokyo station (Tokyo's main railway station) and the imperial palace.

Christian Ghion has focused on basic materials such as blonde wood and white Corian, which are visually very light and revived by fresh and fruity colour such as mandarine and almond green. As usual, he has translated the wishes of the French chef into materials and colour: balanced volumes, simple lines and refined elegance. Through its large plate glass windows, the space is continually bathed in radiant light: distinction that feels as natural as breathing...

设计师克里斯提安-吉翁近来刚完成了一个新的餐饮空间设计,于2007年秋季在东京开幕,名为:"PG咖啡"(皮埃尔-贾内咖啡馆)。设计师保持与名厨皮埃尔-贾内咖啡合作的一贯性,固守两人共通的创作理念。

"PG咖啡"是个很日式风格的咖啡馆,供应点心、浓汤、沙拉和冰淇淋。店面的位置优越,介于东京站(东京最大火车站)和皇宫之间。

克里斯提安-吉翁将焦点投注在基本材质的运用上,如金黄色的木材和白色可丽耐面料,这些材质在视觉上甚为轻盈,在此更以清新微酸的水果色彩(橘色与杏仁绿)来加以强化。如同设计师一贯的风格,他将法国名厨的期望与企图用材质与色彩诠释出来,呈现出均衡的体量、简约纯净的线条和含蓄幽雅的空间气质。这整个由大片落地窗构成的空间,整日沉浸在灿烂绚丽的阳光里,使高贵雅致的空间与轻松舒缓的氛围得到巧妙合韵的搭配。

restaurant directory

餐厅资讯

15CENT15
Hôtel Marignan Champs-Elysées
12 rue de Marignan
75008 Paris France
www.hotelmarignan.fr
reservation@hotelmarignan.fr
T +33 1 40 76 34 56
F +33 1 40 76 34 54

ANGELINA
226 rue de Rivoli
75001 Paris France
T +33 1 42 60 82 00
F +33 1 53 45 83 89

BABOTO
12 rue de la Ferronnerie
75001 Paris France
www.baboto.com
baboto@baboto.com
T +33 1 40 41 11 41
T +33 1 40 41 11 41

BARAMAKI
Printemps Haussmann
64 boulevard Haussmann
75009 Paris France
www.printemps.com
T +33 1 42 82 49 03

BAR DU PORT
7 quai de Suffren
83990 Saint-Tropez France
www.domainetropez.com
info@domainetropez.com
T +33 4 94 97 00 54

BENOIT
La porte Aoyama 10F
5-51-8 Jingumae, Shibuya-ku
Tokyo Japan
www.benoit-tokyo.com
info@benoit.co.jp
T +03 6419 4181
F +03 6419 4185

BERTIE
6 rue Edouard VII
75009 Paris France
www.bertie.fr
bertie1@hotmail.fr
T +33 1 53 05 50 55
F +33 1 53 05 95 56

BLACK CALAVADOS
40 avenue Pierre 1er de Serbie
75008 Paris France
www.blackcalavados.com
info@bc-paris.fr
T +33 1 47 20 77 77

DJOON
22-24 boulevard Vincent Auriol
75013 Paris France
www.djoon.fr
contact@djoon.fr
T +33 1 45 70 83 49
F +33 1 45 70 83 57

EAT ME
38 rue Léopold Belland
75002 Paris France
www.eatme.fr
eatme@eatme.fr
T +33 1 42 36 18 28

FAUCHON
24-26 place de la Madeleine
75008 Paris France
www.fauchon.com
T +33 1 70 39 38 00

FLUNCH
Immeuble Péricentre
Boulevard Van Gogh
59650 Villeneuve d'Ascq France
www.flunch.fr
T +33 3 20 43 59 59
F +33 3 20 43 55 40

GILT
New York Palace Hotel
455 Madison Avenue
New York, NY 10022 USA
T +1 212 891 8100
www.giltnewyork.com

GUILO GUILO
8 rue Garreau
75018 Paris France
T +33 1 42 54 23 92

HÄAGEN DAZS
49/51 avenue des Champs-Elysées
75008 Paris France
www.haagen-dazs.fr
T +33 1 53 77 68 68

HANAWA
26 rue Bayard
75008 Paris France
www.kinugawa-hanawa.com
kinugawa-hanawa@orange.fr
T +33 1 56 62 70 70
F +33 1 56 62 70 71

SHAMWARI
JIVA HILL PARK HOTEL
Route d'Harée
01170 Crozet France
www.jivahill.com
shamwari@jivahill.com
T +33 4 50 28 48 47
F +33 4 50 28 48 49

KAISEKI BENTO
Le Rendez-vous Toyota
79 avenue des Champs-Elysées
75008 Paris France
www.kaisekibento.com
T +33 1 56 89 29 83

L'AUBERGE DE L'ILL
2 rue de Collonges au Mont d'Or
68970 Illhaeusern France
www.auberge-de-l-ill.com
aubergedelill@aubergedelill.com
T +33 3 89 71 89 00
F +33 3 89 71 82 83

LA PATISSERIE PIERRE GAGNAIRE
4F Takashimaya Shinjyyuku Store
5-24-2 Sendagaya, Shibuya-ku
Tokyo Japan
www.pierre-gagnaire.jp
T +03 5361 1977
F +03 5926 4171

LE BOSQUET
46 avenue Bosquet
75007 Paris France
bosquet@wanadoo.fr
T +33 1 45 51 38 13

LE BOUDOIR
14 place Jules Ferry
69006 Lyon France
www.leboudoir.fr
leboudoir1@wanadoo.fr
T +33 4 72 74 04 41

L'ECLAIREUR
10 rue Boissy d'Anglas
75008 Paris France
www.leclaireur.com
bar@leclaireur.com
T +33 1 53 43 09 99

L'EPI DUPIN
11 rue Dupin
75006 Paris France
www.epidupin.com
contact@epidupin.com
T +33 1 42 22 64 56

LE FIRST RESTAURANT BOUDOIR PARIS
234 rue de Rivoli et 3 rue de Castiglione
75001 Paris France
www.lefirstrestaurant.com
le.first@westin.com
T +33 1 44 77 10 40

LES 2 FRERES
4 avenue Reine Astrid
13100 Aix-en-Provence France
www.les2freres.com
les-deuxfreres@wanadoo.fr
T +33 4 42 27 90 32

LES GRANDES MARCHES
6 place de la Bastille
75012 Paris France
www.groupe-bertrand.com
grandes-marches@groupe-bertrand.com
T +33 1 43 42 90 32
F +33 1 43 44 80 02

LE JULES VERNE
Tour Eiffel
Avenue Gustave Eiffel
75007 Paris France
www.lejulesverne-paris.com
T +33 1 45 55 61 44

LE LADUREE BAR
21 rue Bonaparte
75006 Paris France
www.laduree-boutique.com
serviceclient@laduree-boutique.com
T +33 1 44 07 64 87
F +33 1 44 07 64 93

LE RESTAURANT 1947
Hôtel Cheval Blanc
Jardin Alpin
73120 Courchevel 1850 France
www.chevalblanc.com
info@chevalblanc.com
T +33 4 79 00 50 50
F +33 4 79 00 50 51

LE RESTAURANT
13 rue des Beaux-arts
75006 Paris France
www.l-hotel.com
stay@l-hotel.com
T +33 1 44 41 99 00
F +33 1 43 29 90 65

LE SAUT DU LOUP
Musée des Arts Décoratifs
107 rue de Rivoli
75001 Paris France
www.lesautduloup.com
audrey.fouchard@groupe-sdw.com
T +33 1 42 25 49 55
F +33 1 42 25 49 60

LE TARMAC
33 rue de Lyon
75012 Paris France
www.tarmac-paris.com
contact@tarmac-paris.com
T +33 1 43 41 97 70

LA TASSEE
20 rue de la Charité
69002 Lyon France
www.latassee.fr
jpbmessimy@latassee.fr
T +33 4 72 77 79 00
F +33 4 72 40 05 91

LE TELEGRAPHE
41 rue de Lille
75007 Paris France
www.restaurantletelegraphe.com
info@restaurantletelegraphe.com
T +33 1 58 62 10 08
F +33 1 58 62 10 09

MAMA SHELTER
109 rue de Bagnolet
75020 Paris France
www.mamashelter.com
paris@mamashelter.com
T +33 1 43 48 48 48
F +33 1 43 48 49 49

MARKET
15 avenue Matignon
75008 Paris France
www.jean-georges.com
prmarketsa@aol.com
T +33 1 56 43 40 90
F +33 1 56 43 40 92

MINI PALAIS
Grand Palais
Avenue Winston Churchill
75008 Paris France
www.minipalais.com
info@minipalais.com
T +33 1 42 56 42 42

MOPH
Parco part1 - 1F
15-1 Udagawa-cho, Shibuya-ku
Tokyo Japan
www.moph.jp
contact@moph.jp
T/F +03 5456 8244

MYBERRY
25 rue Vieille du Temple
75004 Paris France
www.myberry.eu
info@myberry.eu
T +33 1 42 74 54 48

NABULIONE
40 avenue Duquesne
75007 Paris France
www.nabulione.com
restaurantnabulione@gmail.com
T +33 1 53 86 09 09
F +33 1 53 86 09 10

OSMOSE
31 avenue de Versailles
75016 Paris France
www.osmose-paris.com
contact@osmose-paris.com
T +33 1 45 20 74 12

PG CAFE
Shin-Marunouchi Bldg.
1-5-1 Marunouchi, Chiyoda-ku
Tokyo Japan
www.pierre-gagnaire.jp
T +03 5224 3455

LE PRE CATELAN
Route de Suresnes, Bois de Boulogne
75016 Paris France
www.precatelanparis.com
leprecatelan-restaurant@lenotre.fr
T +33 1 44 14 41 14
F +33 1 45 24 43 25

REFLETS PAR PIERRE GAGNAIRE
InterContinental Dubai Festival City
PO Box 45777 Dubai UAE
reflets.restaurant@ichdfc.ae
www.ichotelsgroup.com
T +971 4 701 1199
F +971 4 232 9095

ROYCE
3 rue des Saussaies
75008 Paris France
www.royce-events.com
contact@royce-events.com
T +33 1 43 12 82 00
F +33 1 43 12 82 05

SHAKE EAT
17 rue de Choiseul
75002 Paris France
www.shake-eat.fr
T/F +33 1 40 07 90 64

SUR UN ARBRE PERCHE
1 rue du Quatre Septembre
75002 Paris France
www.surunarbreperche.com
contact@surunarbreperche.com
T +33 1 42 96 97 01

TOKYO EAT
13 avenue du Président Wilson
75116 Paris France
www.tokyoeat.com
T +33 1 47 20 00 29
F +33 1 47 20 05 62

TOUSTEM
www.helenedarroze.com

designer directory

设计师资讯

Anégil
Architecture intérieure & design
15 rue de Picardie
75003 Paris France
www.anegil.com
anegil@anegil.com
T +33 1 48 06 05 66
F +33 1 48 04 04 37

Afshin Assadian
22 boulevard Vincent Auriol
75013 Paris France
www.djoon.fr
afshin@djoon.fr
T +33 1 45 70 83 49
F +33 1 45 70 83 57

Atelier FB
28 rue du Sentier
75002 Paris France
www.atelier-fb.com
fb@atelier-fb.com
T +33 1 40 06 96 60
F +33 1 42 68 07 54

Christian Biecher
14 rue Crespin du Gast
75011 Paris France
www.biecher.com
info@biecher.com
T +33 1 49 29 69 39
F +33 1 49 29 69 30

Alexandre de Betak's PA
2 rue des Haudriettes
75003 Paris France
www.bureaubetak.com
bbcontact@bureaubetak.com

Philippe Boisselier
14 rue de Rivoli
75004 Paris France
www.philippeboisselier.com
boisselier.philippe@wanadoo.fr
T +33 1 42 78 11 82
F +33 1 42 71 24 26

Code Déco
11 rue Tronchet
75008 Paris France
www.code-deco.com
codedeco@wanadoo.fr
T +33 1 40 06 99 12
F +33 1 40 06 99 08

Claudio Colucci
59 rue Meslay
75003 Paris France
www.colucci-design.com
contact@colucci-design.com
T +33 1 44 73 00 20
F +33 1 42 77 32 29

Matali Crasset
26 rue du Buisson Saint-Louis
75010 Paris France
www.matalicrasset.com
matali.crasset@wanadoo.fr
T +33 1 42 40 99 89
F +33 1 42 40 99 98

Dragon Rouge
32 rue Pagès BP 83
92153 Suresnes Cedex France
www.dragonrouge.com
info@dragonrouge.com
T +33 1 46 97 50 00
F +33 1 41 72 05 03

Olivier Gagnère & Associés
47 boulevard Saint-Jacques
75014 Paris France
www.gagnere.net
olivier@gagnere.net
T/F +33 1 45 80 79 67

Jacques Garcia
212 rue de Rivoli
75001 Paris France
www.decojacquesgarcia.com
cwhite@decojacquesgarcia.com
T +33 1 42 97 48 70
F +33 1 42 97 48 73

Christian Ghion
156 rue Oberkampf
75011 Paris France
www.christianghion.com
ghion@christianghion.com
T +33 1 49 29 06 90
F +33 1 49 29 06 89

Frédérique Gormand
12 rue Dupetit-Thouars
75003 Paris France
gormandf@yahoo.fr
T +33 1 40 29 48 39

Olivier Gossart
9 rue de Valadon
75007 Paris France
www.o-gossart.com
contact@o-gossart.com
T +33 1 42 60 14 80
F +33 1 42 60 14 85

Armand & Martine Hadida
Piero Fornasetti
L'Eclaireur Faubourg Saint-Honoré
10 rue Boissy d'Anglas
75008 Paris France
www.leclaireur.com
T +33 1 53 43 03 70

Idoine
44 rue du Château d'Eau
75010 Paris France
www.groupeidoine.com
contact@groupeidoine.com
T +33 1 42 06 10 10
F +33 1 42 06 24 30

Jouin Manku
8 passage de la Bonne Graine
75011 Paris France
www.patrickjouin.com
agence@jouinmanku.com
T +33 1 55 28 89 20
F +33 1 58 30 60 70

L'Agence (Interior Design)
Régis Conseil & Michael Malapert
1 boulevard Saint-Denis
75003 Paris France
www.lagence-paris.com
contact@lagence-paris.com
T/F +33 1 42 22 81 25

Le Gall & Polianoff
3 rue Séguier
75006 Paris France
www.legallpolianoff.com
patrick.polianoff@legallpolianoff.com
gilles.le-gall@legallpolianoff.com
T +33 1 45 45 14 15
F +33 1 45 45 26 27

Architecte d'intérieur Christian Liaigre
122 rue de Grenelle
75007 Paris France
www.christian-liaigre.fr
sales@christian-liaigre.fr
T +33 1 70 08 65 20

Xavier Luvison & Jean-Christophe Sabarthès
260 avenue du 24 avril 1915
13012 Marseille France
www.ls-archi.com
contact@ls-archi.com
T +33 4 91 93 82 28
F +33 4 91 93 82 27

Sybille de Margerie
9 rue Emile Allez
75017 Paris France
www.smdesign.fr
smdesign@smdesign.fr
T +33 1 40 55 70 70
F +33 1 40 55 70 71

Stéphane Maupin & Nicolas Hugon
1 rue Sarasate
75015 Paris France
www.stephanemaupin.com
contact@stephanemaupin.com
T/F +33 1 44 26 06 25

Jean-Philippe Nuel
9 boulevard de la Marne
94130 Nogent-sur-Marne France
www.jeanphilippenuel.com
jpn@jeanphilippenuel.com
T +33 1 45 14 12 10
F +33 1 48 77 26 92

Gérard Olouman
3 rue des Saussaies
75008 Paris France
www.royce-events.com
contact@royce-events.com
T +33 1 43 12 82 00
F +33 1 43 12 82 05

Ora-Ïto
58 rue Charlot
75003 Paris France
www.ora-ito.com
info@ora-ito.com
T +33 1 42 46 00 09
F +33 1 42 46 03 09

Jean-Patrice Pham
13 avenue Marie Juliette
92250 La Garenne-Colombes France
www.cabinetpham.com
contact@cabinetpham.com
T/F +33 1 47 82 48 05
P +33 6 24 55 46 19

Donatelle Piana & Philippe Batifoulier
Agence DP id
39 rue de l'Université
69007 Lyon France
donatelle.p@wanadoo.fr
philippe.batifoulier@numericable.fr
T +33 4 78 95 30 16

Christophe Pillet
29 passage Dubail
75010 Paris France
www.christophepillet.com
info@christophepillet.com
T +33 1 58 36 46 31
F +33 1 42 25 01 25

Ralston & Bau
10 rue André Barsacq
75018 Paris France
P +33 6 09 20 18 98

Ralston & Bau, Transplant
6963 Dale i Sunnfjord Norway
www.ralstonbau.com
about@ralstonbau.com
T +47 57 73 55 97

Pierre-Yves Rochon
9 avenue Matignon
75008 Paris France
www.pyr-sa.com
pyrparis@pyr-sa.com
T +33 1 44 95 84 84
F +33 1 44 95 84 70

Roxane Rodriguez
18 rue de Seine
75006 Paris France
www.roxanerodriguez.fr
secretariat@roxanerodriguez.fr
T +33 1 44 32 11 10
F +33 1 43 25 41 91

Saguez & Partners
Manufacture Design
14 rue Palouzié
93400 Saint-Ouen France
www.saguez-and-partners.com
info@saguez-and-partners.com
T +33 1 41 66 64 00
F +33 1 41 66 64 01

SCP d'architectes
Denis Boyer-Gibaud François Percheron Antoine Assus
36 avenue de Lodève
34070 Montpellier France
www.boyer-percheron-assus.com
architecture-boyer-percheron@wanadoo.fr
T +33 4 67 41 49 40
F +33 4 67 04 12 75

Valérie Serin
146 rue de la Pompe
75116 Paris France
www.valerieserin.com
sv@valerieserin.com
P +33 6 12 89 57 18
F +33 1 56 28 01 35

Philippe Stark / Ubik
18-20 rue du Faubourg du Temple
75011 Paris France
www.starck.com
T +33 1 48 07 54 54
F +33 1 48 07 54 64

Christophe Vendel
62 rue Jean-Baptiste Pigalle
75009 Paris France
www.questionsdinterieur.com
christophe.vendel@gmail.com

图书在版编目（ＣＩＰ）数据

新法国餐厅设计／法国亦西文化编著.——沈阳：辽宁科学技术出版社，2009.4
　ISBN 978-7-5381-5910-3

　I.新… II.法… III.餐厅－室内设计－法国－图集
IV.TU247.3-64

中国版本图书馆CIP数据核字（2009）第031012号

出版发行：辽宁科学技术出版社
　　　　　（地址：沈阳市和平区十一纬路29号　邮编：110003）
印　刷　者：利丰雅高印刷（深圳）有限公司
经　销　者：各地新华书店
幅面尺寸：240mm×320mm
印　　张：44.5
插　　页：4
字　　数：150千字
印　　数：1~3000
出版时间：2009年4月第1版
印刷时间：2009年4月第1次印刷
责任编辑：陈慈良
封面设计：亦　西
版式设计：亦　西
责任校对：周　文

书　　号：ISBN 978-7-5381-5910-3
定　　价：298.00元

联系电话：024-23284360
邮购热线：024-23284502
E-mail: lkzzb@mail.lnpgc.com.cn
http://www.lnkj.com.cn